P9-DUP-360

METHODS IN RADIOIMMUNOASSAY, TOXICOLOGY, AND RELATED AREAS

PROGRESS IN ANALYTICAL CHEMISTRY

Based upon the Eastern Analytical Symposia

Series Editors:
Ivor L. Simmons
M&T Chemicals, Inc., Rahway, New Jersey
and Galen W. Ewing
Seton Hall University, South Orange, New Jersey

Volume 1
H. van Olphen and W. Parrish
X-RAY AND ELECTRON METHODS OF ANALYSIS
Selected papers from the 1966 Eastern Analytical Symposium

Volume 2
E. M. Murt and W. G. Guldner
PHYSICAL MEASUREMENT AND ANALYSIS OF THIN FILMS
Selected papers from the 1967 Eastern Analytical Symposium

Volume 3
K. M. Earle and A. J. Tousimis
X-RAY AND ELECTRON PROBE ANALYSIS IN BIOMEDICAL RESEARCH
Selected papers from the 1967 Eastern Analytical Symposium

Volume 4
C. H. Orr and J. A. Norris
COMPUTERS IN ANALYTICAL CHEMISTRY
Selected papers from the 1968 Eastern Analytical Symposium

Volume 5
S. Ahuja, E. M. Cohen, T. J. Kneip, J. L. Lambert, and G. Zweig
CHEMICAL ANALYSIS OF THE ENVIRONMENT AND OTHER MODERN TECHNIQUES
Selected papers from the 1971 Eastern Analytical Symposium

Volume 6
Ivor L. Simmons and Galen W. Ewing
APPLICATIONS OF THE NEWER TECHNIQUES OF ANALYSIS
Selected papers from the 1972 Eastern Analytical Symposium

Volume 7
Ivor L. Simmons and Galen W. Ewing
METHODS IN RADIOIMMUNOASSAY, TOXICOLOGY, AND RELATED AREAS
Selected papers from the 1973 Eastern Analytical Symposium

PROGRESS IN ANALYTICAL CHEMISTRY
VOLUME 7

METHODS IN RADIOIMMUNOASSAY, TOXICOLOGY, AND RELATED AREAS

Edited by
Ivor L. Simmons
M&T Chemicals, Inc.
Rahway, New Jersey

and

Galen W. Ewing
Seton Hall University
South Orange, New Jersey

PLENUM PRESS • NEW YORK AND LONDON

Library of Congress Cataloging in Publication Data

Eastern Analytical Symposium, Atlantic City, 1973.
Methods in radioimmunoassay, toxicology, and related areas.

(Progress in analytical chemistry; v. 7)
Includes bibliographies and index.
1. Chemistry, Clinical—Congresses. 2. Radioimmunoassay—Congresses. 3. Toxicology—Congresses. I. Simmons, Ivor L., ed. II. Ewing, Galen Wood, 1914- ed. III. Title.
RB40.E18 1973 616.07'9 74-23819
ISBN 0-306-39307-7

A Division of Plenum Publishing Corporation
227 West 17th Street, New York, N.Y. 10011

United Kingdom edition published by Plenum Press, London
A Division of Plenum Publishing Company, Ltd.
4a Lower John Street, London, W1R 3PD, England

Printed in the United States of America

PREFACE

The Fifteenth Eastern Analytical Symposium and the Twelfth National Meeting of the Society for Applied Spectroscopy were held jointly in New York, November 14-16, 1973. In cooperation with the Detroit Analytical Chemists' Association, the Anachem Award in Analytical Chemistry was presented at this meeting to Dr. Rosalyn S. Yalow, the co-inventor (with Dr. S. Berson) of the radioimmunoassay technique.

It is our privilege to include in this volume Dr. Yalow's award address, a review of radioimmunoassay, together with a series of three papers on related subjects, respectively by Dr. Juan M. Saavedra, Drs. Claude E. Arnaud and H. Bryan Brewer, Jr., and Dr. Sheldon P. Rothenberg. This session was organized and chaired by Dr. D. S. McCann.

Following this are three papers on analytical toxicology in clinical chemistry, by Dr. R. A. Scala, Dr. Charles L. Winek, and Dr. James E. Gibson, a session also chaired by Dr. McCann. Finally appear two papers on the general subject of analytical developments at the FDA, by Dr. Albert L. Woodson and Dr. Walter Holak, from a session under the chairmanship of Dr. S. Walters.

All of these contributions, written by experts in their respective fields, make worthy additions to the continuing literature of analytical chemistry.

Ivor L. Simmons

Galen W. Ewing

CONTENTS

RADIOIMMUNOASSAY: ITS PAST, PRESENT, AND POTENTIAL

Rosalyn S. Yalow, Ph.D.
Senior Medical Investigator
Project #9678-01
Veterans Administration Hospital
Bronx, New York 10468
Dept. of Med., Mt. Sinai School of Med.
The City University of New York

I am appreciative of the honor bestowed on me by receiving the Anachem Award for distinguished service to analytical chemistry. It is perhaps worth noting that this is the first time that the award has been presented for investigations in bioanalytic methodology and the first time that the recipient was trained as a nuclear physicist. In these days when there is a tendency for increasing support for contract research, for crash programs for instant solutions to problems such as cancer or the energy crisis, it is important to remember that while these programs are designed to exploit most efficiently previous scientific breakthroughs, revolutionary ideas continue to arise unpredictably from individual scientists or small groups and the benefits derived therefrom may well transcend the borders of any particular scientific discipline.

Today I would like to share with you the history of the development of radioimmunoassay and to discuss some aspects of the applicability of this bioanalytical tool in clinical medicine, physiology and biochemistry.

In the early 1950's the late Dr. Solomon A. Berson and I began our investigative careers in the application of radioisotopes in thyroid physiology and diagnosis and in blood volume determinations with radioisotope-labeled red cells and serum proteins. We soon turned our attention to the in-vivo distribution and metabolism of ^{131}I-labeled proteins (1-3) and, from this, extension to studies with ^{131}I-labeled peptidal hormones appeared promising.

At that time insulin was the hormone most readily available in a highly purified form. Dr. I. Arthur Mirsky had just proposed a hypothesis (4) that maturity-onset diabetes might not be due to a deficiency of insulin secretion but rather to abnormally rapid degradation of insulin by hepatic insulinase. We therefore began to investigate the distribution and metabolism of ^{131}I-labeled insulin administered intravenously to non-diabetic and diabetic subjects (5). Contrary to the predictions of the Mirsky hypothesis, we observed that radioactive insulin generally disappeared more slowly from the plasma of diabetic subjects than from the plasma of non-diabetic controls (Fig. 1).

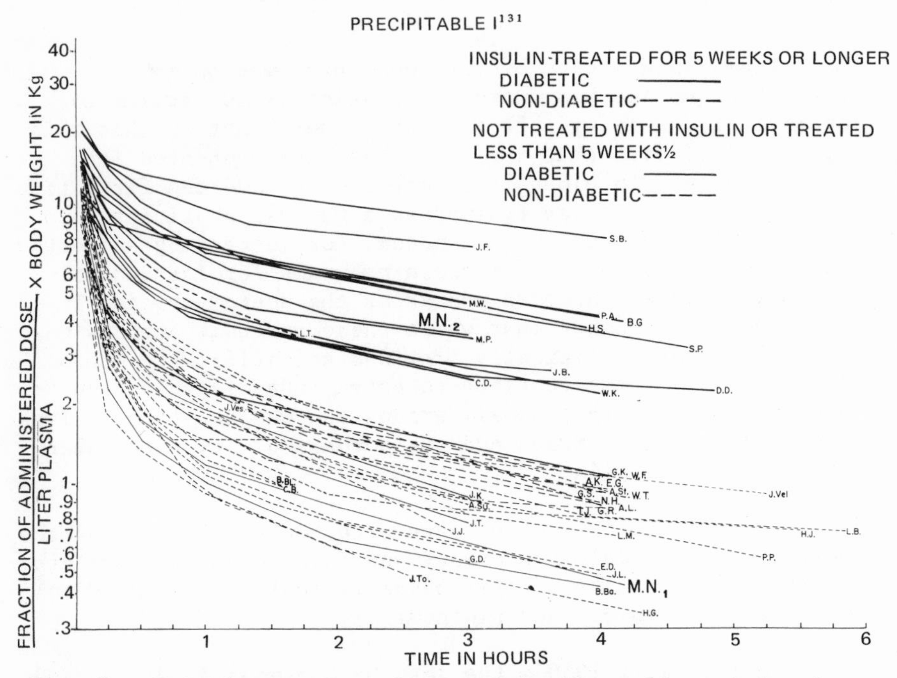

Fig. 1 - Trichloracetic acid precipitable radioactivity in plasma as a function of time following intravenous administration of ^{131}I-insulin in diabetic and non-diabetic subjects. M. N.$_1$ - after treatment with insulin for 2 1/2 weeks; M. N.$_2$ - after treatment with insulin for 4 1/2 months. (Reproduced from ref. 5)

However, one should remember that these studies were performed in the early 1950's, before the era of oral hypoglycemic agents, such as the sulfonylureas, and that most diabetic subjects had received insulin therapy. We soon appreciated that the difference between the two groups was not due to diabetes per se but rather that those patients from whose plasma the insulin disappeared slowly had a previous history of insulin therapy.

We suspected that the retarded rate of insulin disappearance was caused by binding of labeled insulin to insulin antibodies which had developed in response to administration of exogenous insulin. However classic immunologic techniques were not adequate for the detection of the low concentrations of antibody presumed to be present. We therefore developed new and highly sensitive techniques employing radioiodine-labeled insulin for the demonstration of binding to antibody in vitro. These included paper electrophoresis and chromato-electrophoresis (Fig. 2). In common with many other peptidal

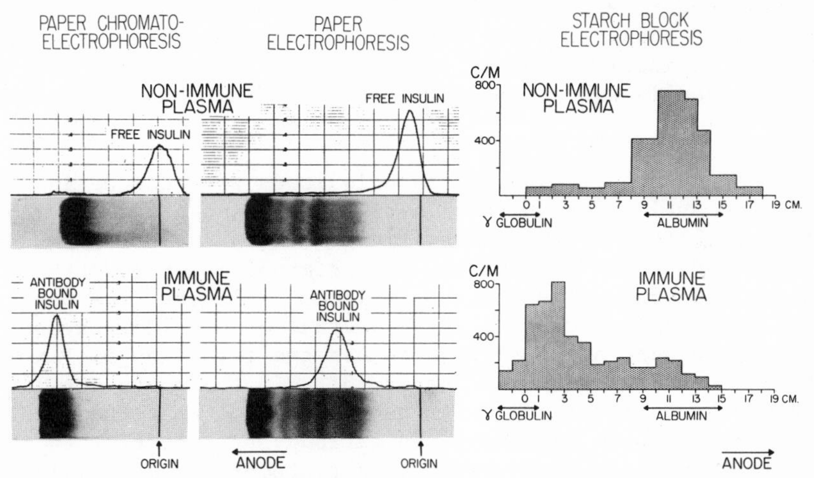

Fig. 2 - ^{131}I-insulin was added to the plasmas of insulin-treated (bottom) and untreated (top) human subjects and the mixtures were applied to a starch block (right) or to paper strips (middle) for electrophoresis or to paper strips for hydrodynamic flow chromatography combined with electrophoresis (left). After completion of electrophoresis, segments were cut out of the starch block for assay of radioactivity and the paper strips were assayed in an automatic strip counter. (Reproduced from Berson and Yalow: The Harvey Lectures, Series 62, Acad. Press, N.Y. 1968, p. 107.)

hormones, insulin in its free state tends to absorb firmly to paper. In contrast, it does not do so when bound to antibody. Therefore, if an incubated mixture of labeled insulin and plasma is applied to a strip of filter paper for electrophoresis or chromatoelectrophoresis, the labeled insulin in the plasmas of untreated subjects remains at the site of application, i.e., the origin, whereas, in the plasma of treated patients in whom antibody had developed, the insulin bound to the antibody moves between the gamma and the beta globulins. On chromatoelectrophoresis, the serum proteins do not separate significantly from each other, but, within a matter of only 20-30 minutes, all of the serum proteins carrying the insulin-antibody complexes have migrated away from the origin. On starch block electrophoresis, labeled insulin in the immune plasma remains with the inter β-γ globulins close to site of application and the free insulin which is not absorbed to the starch has an electrophoretic mobility almost that of albumin (Fig. 2). During ultracentrifugation, labeled insulin, having a molecular weight of about 6000, sedimented more slowly than serum albumin in the plasma of untreated subjects, but sedimented with the globulins in the plasma of insulin-treated subjects (5). Binding of insulin to γ-globulin was demonstrable on salt fractionation as well (5).

The demonstration of the ubiquitous presence of insulin-binding antibodies in insulin-treated subjects was not readily acceptable to the scientific community 20 years ago. It is perhaps worth noting that the original paper describing these finding was rejected by Science. Although the paper was eventually accepted by the Journal of Clinical Investigation we were not permitted to use the term 'insulin antibody' in the title of the paper (5). In the paper itself we were required to document our conclusion that the insulin-binding globulin was indeed an antibody by referring to the definition of antibody given in a textbook of bacteriology and immunity (6).

In this early work (5) we also observed that the binding of labeled insulin is a quantitative function of the amount of insulin present when the antibody concentration is kept fixed. It was this observation that provided the basis of the radioimmunoassay of plasma insulin. However several years were to pass from the presentation of the immunoassay principle (5, 7) to its practical application to the measurement of plasma insulin in man. During that

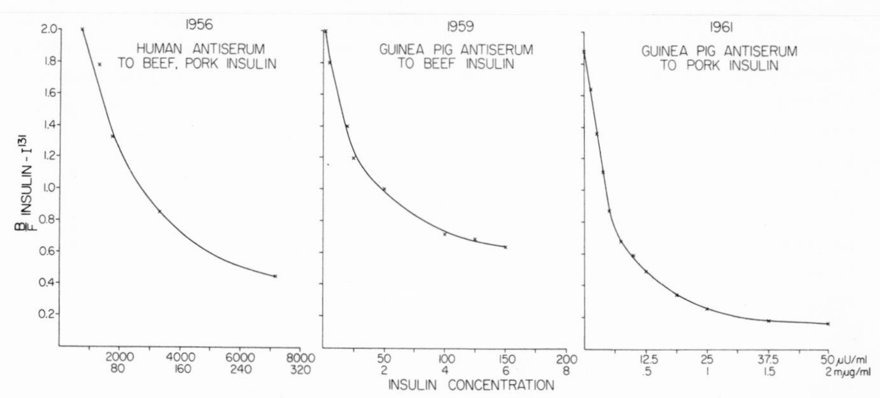

Fig. 3 - Standard curves for the measurement of insulin during the period 1956 to 1961. The human antiserum (left) could be used to detect 1000 μU beef insulin/ml. More than a 10 fold higher concentration of human insulin were required to effect the same reduction in B/F. The guinea pig antiserum to beef insulin (middle) could be used to detect 10 μU human insulin/ml. The guinea pig antiserum to pork insulin (right) had a 5-10 fold greater sensitivity for the detection of human insulin. This improvement in sensitivity was required to make radioimmunoassay of human plasma insulin possible since the average insulin concentration after an overnight fast is less than 20 μU/ml.

time we studied the quantitative aspects of the reaction between insulin and antibody (8), evaluated the species-specificity of the available antisera (9) and measured the disappearance of exogenous beef insulin administered to rabbits (10). These investigations provided the theoretical and experimental basis for the measurement of insulin in unextracted human plasma (11, 12). As shown in Fig. 3, the

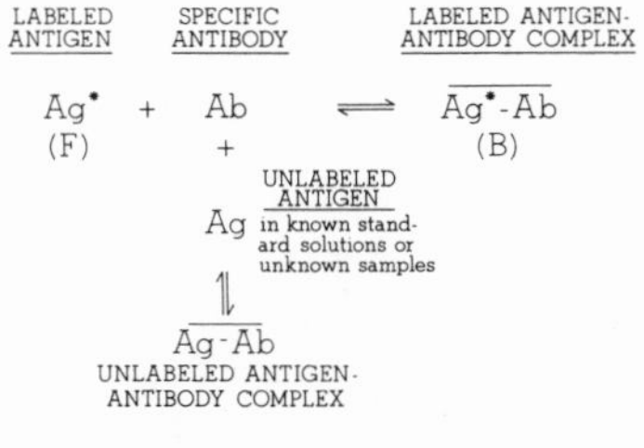

LABELED ANTIGEN | SPECIFIC ANTIBODY | LABELED ANTIGEN-ANTIBODY COMPLEX

$$\mathrm{Ag^{*}}\ (F) + \mathrm{Ab} \rightleftharpoons \overline{\mathrm{Ag^{*}\text{-}Ab}}\ (B)$$

\+

UNLABELED ANTIGEN: Ag in known standard solutions or unknown samples

$$\rightleftharpoons$$

$\overline{\mathrm{Ag\text{-}Ab}}$
UNLABELED ANTIGEN-ANTIBODY COMPLEX

Fig. 4 - Competing reactions that form the basis of the radioimmunoassay.

PEPTIDE HORMONES

PITUITARY HORMONES
- Growth hormone
- Adrenocorticotropic hormone (ACTH)
- Melanocyte stimulating hormone (MSH)
  - α-MSH
  - β-MSH
- Glycoproteins
  - Thyroid stimulating hormone (TSH)
  - Follicle stimulating hormone (FSH)
  - Luteinizing hormone (LH)
- Prolactin
- Lipotropin (LPH)
- Vasopressin
- Oxytocin

CHORIONIC HORMONES
- Human chorionic gonadotropin (HCG)
- Human chorionic somatomammotropin (HCS)

PANCREATIC HORMONES
- Insulin
- Proinsulin
- C-peptide
- Glucagon

CALCITROPIC HORMONES
- Parathyroid hormone (PTH)
- Calcitonin (CT)

GASTROINTESTINAL HORMONES
- Gastrin
- Secretin
- Cholecystokinin-pancreozymin (CCK-PZ)
- Enteroglucagon
- Vasoactive intestinal polypeptide (VIP)
- Gastric inhibitory polypeptide (GIP)

VASOACTIVE TISSUE HORMONES
- Angiotensins
- Bradykinins

HYPOTHALAMIC RELEASING FACTORS
- Thyrotropin releasing factor (TRF)
- Gonadotropin releasing factor (GnRF)

NON-PEPTIDAL HORMONES

THYROIDAL HORMONES
- Triiodothyronine
- Thyroxine

PROSTAGLANDINS

STEROIDS
- Aldosterone
- Corticosteroids
- Estrogens
- Androgens
- Progesterones

NON-HORMONAL SUBSTANCES

DRUGS
- Digoxin
  - Digitoxin
- Morphine
- LSD
- Barbiturates

CYCLIC NUCLEOTIDES
- cAMP
- cGMP
- cIMP
- cUMP

ENZYMES
- C_1 esterase
- Fructose 1,6 diphosphatase

VIRUS
- Australia antigen (Hepatitis B antigen)

TUMOR ANTIGENS
- Carcinoembryonic antigen
- α-Fetoprotein

SERUM PROTEINS
- Thyroxin binding globulin
- IgG, IgE
- Properdin
- Anti-Rh antibodies

OTHER
- Intrinsic factor
- Rheumatoid factor
- Folic acid
- Neurophysin
- Staphylococcal
  - β-Enterotoxin

Fig. 5 - Partial listing of peptidal and non-peptidal hormones and other substances measured by radioimmunoassay.

principal improvement in sensitivity for the detection of human insulin was made possible by the use of guinea pig anti-bovine insulin sera instead of human anti-beef, pork insulin sera (13). Further improvement in sensitivity was effected by the use of guinea pig anti-porcine insulin sera (14).

Radioimmunoassay is simple in principle. It is summarized in the competing reactions shown in Fig. 4. The concentration of the unknown unlabeled antigen is obtained by comparing its inhibitory effect on the binding of radioactively labeled antigen to specific antibody with the inhibitory effect of known standards. The validity of a radioimmunoassay procedure is dependent on identical immunologic behavior of antigen in unknown samples with the antigen in known standards. There is no requirement for identical immunologic or biologic behavior of labeled and unlabeled antigen. Furthermore, as will be considered in some detail later, there is no requirement for standards and unknowns to be identical chemically or to have identical biologic activity. An incomplete listing of substances measured by radioimmunoassay is given in Fig. 5. The rapid rate of growth in this field precludes continuous updating of the listing.

The radioimmunoassay principle can be generalized and extended to non-immune systems to which a more general set of equations (Fig 6) and a more general term, "competitive radioassay" can be applied. Even this term is not completely satisfactory since the principle can be applied in situations in which the marker is other than a radioisotopic tracer. Using competitive radioassay Rothenberg (15) and Barakat and Ekins (16) independently developed assays for serum vitamin B_{12} using intrinsic factor as the specific reactor. Ekins

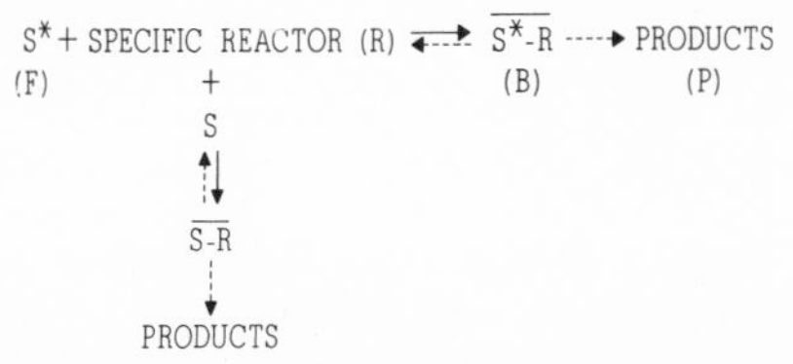

Fig. 6 - Application of competitive radioassay principle to physical or chemical reactions in non-immune systems. Degree of competitive reactions may be determined from changes in concentration of labeled free substrate (S*) or substrate-reactor complex (S*-R) or, as in the case of enzymes, inhibition of product formation.

first reported on the measurement of thyroxine in human plasma using thyroxine-binding globulin as the specific reactor (17). Further development of the thyroxine assay and the application of competitive radioassay to the measurement of plasma cortisol using cortisol binding globulin were carried out by Murphy (18). More recently there has been extensive application of the principle to the measurements of peptidal and non-peptidal hormones using tissue receptor sites as specific reactors (See 19 for review).

Dr. Rothenberg reviews his recent studies with competitive radioassay in a subsequent paper in this symposium.

Before considering more recent work from our laboratory involving radioimmunoassay, let us return to Dr. Mirsky's hypothesis. The first application of the radioimmunoassay of plasma insulin in man was for the determination of the plasma insulin response to a standard oral glucose tolerance test (100 g glucose) in non-diabetic and in adult mild maturity-onset diabetic subjects (Fig. 7) (12). An unexpected

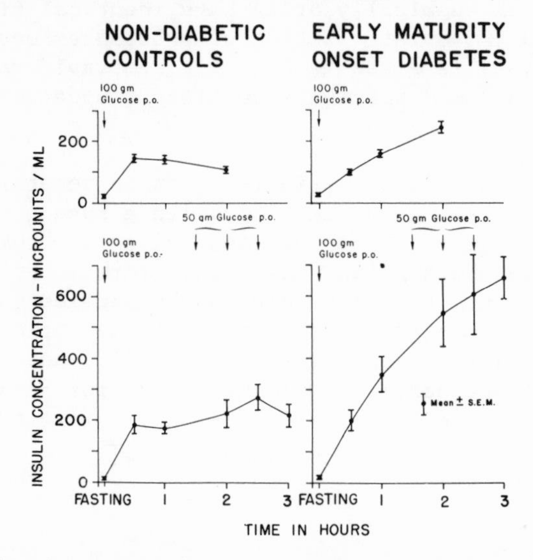

Fig. 7 - (Top) Mean plasma insulin concentration before and during a standard 100 gm oral glucose tolerance test in subjects never treated with insulin (30 non-diabetic subjects and 38 diabetic subjects). (Bottom) Same as above except that additional loads of 50 gm glucose were administered during the test at 1 1/2, 2 and 2 1/2 hours. (4 diabetic subjects and 5 non-diabetic subjects). (Reproduced from Yalow and Berson: Diabetes 10:339, 1961)

discovery was that although the non-diabetic subjects showed, on the average, a more brisk early insulin-secretory response, most of the diabetic patients revealed a capacity to secrete even greater amounts of insulin than would have been sufficient to restore the non-diabetic subjects rapidly to a euglycemic state. That the response of the diabetic in these instances does not represent maximal stimulation of islet tissue is indicated by the very much higher plasma insulin levels that can be achieved in these patients under the stimulus of further heavier glucose loading (Fig. 7) (12). From these investigations we concluded that in the mild maturity-onset diabetic subject absolute insulin deficiency per se is not the cause of hyperglycemia but rather that there is some inherent reason for the reduced responsiveness of the diabetic to apparently adequate amounts of plasma insulin. Now some 14 years later the problem has still not been resolved although, using radioimmunoassay, it is well appreciated that insulin-insensitivity can be associated not only with diabetes but can be induced by obesity, fasting, etc.

Radioimmunoassay has proven useful not only for studies of the physiologic regulation of hormonal secretion, one example of which has just been given, but also in the diagnosis of diseases associated with hormonal excess or hormonal deficiency. Proper interpretation of hormone measurements in these situations requires a clear understanding of the factors involved in the regulation of hormonal secretion. Generally hormonal secretion is stimulated by some departure from the state of biologic "homeostasis" that the hormone is designed to modulate. A model for one such system is shown in Fig. 8. Regulation is effected through the opera-

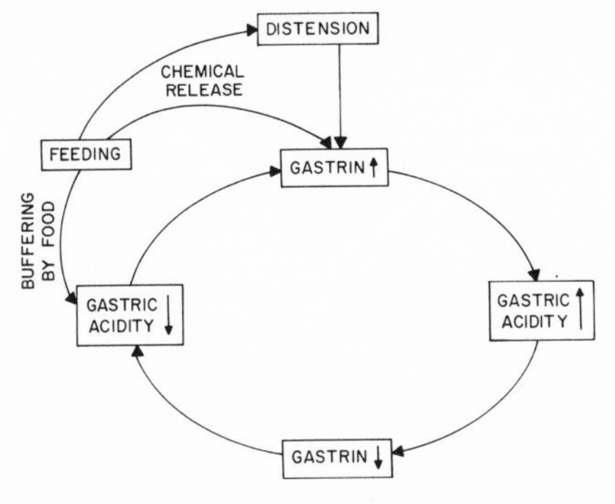

Fig. 8 - Feed-back control loop for gastrin regulation of gastric acidity : effect of feeding.

tion of a feed-back control loop which contains the hormone at one terminus and, at the other, the substance which it regulates and by which it is in turn regulated. Gastrin secretion increases gastric acidity, which in turn suppresses secretion of antral gastrin. Modulation of this system can be effected by a number of factors, perhaps the most important of which is feeding. Feeding promotes gastrin release directly through a chemical effect on antral cells and indirectly through gastric distension and through the buffering action of food (Fig. 8). In Fig. 9 are compared gastrin concentrations in normal subjects, in patients with pernicious anemia (PA) and in patients with Zollinger-Ellison (Z-E) syndrome, who have gastrin-producing tumors. Z-E patients have a very high acid secretion. In these patients the high level of plasma gastrin is inappropriate and demonstrates the failure of the feed-back mechanism either because the tumor is autonomous or because it is generally not located in the stomach and is therefore not bathed by the acid secretion (20). Patients with pernicious anemia have gastric

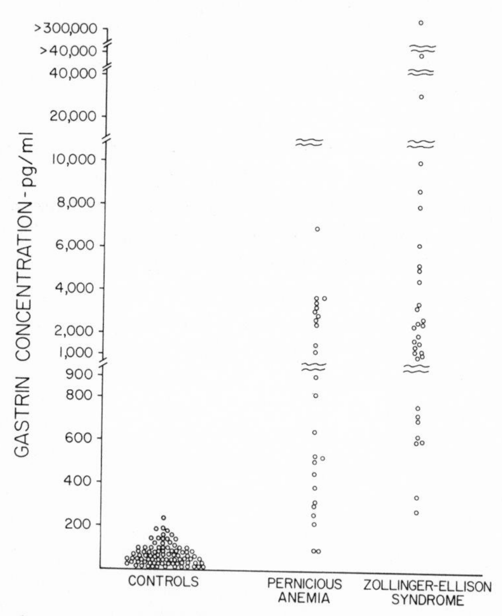

Fig. 9 - Basal plasma gastrin concentrations in normal subjects and in patients with pernicious anemia and Zollinger-Ellison syndrome. (Reproduced from Berson and Yalow: Rhode Island Med. J. 54:501, 1971)

hypoacidity. Since gastric hydrochloric acid normally suppresses gastrin secretion, the continued absence of acid and the repeated stimulation by feeding eventually produces secondary hyperplasia of gastrin-producing cells. The high level of plasma gastrin in pernicious anemia patients is quite appropriate in view of the absence of the inhibitory effect of HCl on the secretion of antral gastrin (20). In pernicious anemia patients the introduction of HCl into the stomach provokes a precipitous fall in plasma gastrin (20).

The radioimmunoassay of plasma gastrin in patients with marked hyperchlorhydria generally provides a valuable diagnostic test for Z-E syndrome since the values in this condition are usually much greater than in the normal or in the usual patient with duodenal ulcer. However over the past several years we have encountered a group of patients with a syndrome we have termed "hypergastrinemic hyperchlorhydria", which resembles Z-E syndrome but which can be distinguished from it. In this syndrome, the plasma gastrin levels are quite high and overlap with those of Z-E patients, the patients may or may not suffer from peptic ulcer and have rates of acid secretion which are generally only modestly elevated and within the range of that found in duodenal ulcer patients rather than in the much higher Z-E range. This syndrome appears to rise from overactivity of the gastrin-secreting cells of the gastrointestinal tract and, as a result, patients with this syndrome manifest an excellent gastrin-secretory response to feeding (Fig. 10) (21). In contrast, the Z-E patient shows no significant elevation of plasma gastrin on feeding (Fig. 10) probably because the long-standing hyperchlorhydria suppresses the gastrin-secreting cells of the antral and duodenal mucosa and because the tumor itself is not in the gastrointestinal tract and is not in contact with food.

Thus in the application of radioimmunoassay to problems of hormonal hypo- or hypersecretion we seldom should rely on a single determination of plasma hormone. Generally, to test for deficiency states, plasma hormonal concentrations should be measured not only in the basal state but also following administration of appropriate physiologic or pharmacologic stimuli. When hypersecretion is suspected and high hormonal levels are observed, one must determine whether the hormonal level is appropriate or inappropriate and whether the hormonal secretion is autonomous or can be modulated by appropriate physiologic or pharmacologic agents. Requirements

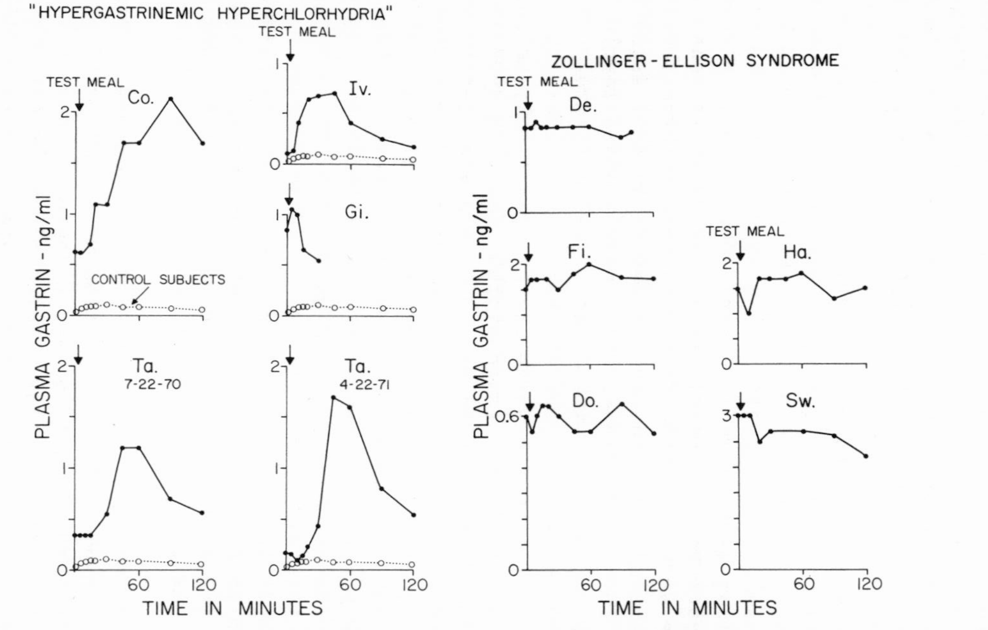

Fig. 10 - Plasma gastrin concentrations in basal state and following test meal (4 oz. orange juice, 2 eggs, 1 piece dry toast): (Left) In gastrin hypersecretors with hyperchlorhydria. Co, Gi and Ta have had duodenal ulcers. Iv has carried a diagnosis of peptic esophagitis but has had neither a typical ulcer history nor roentgen evidence of ulcer; (Right) In patients with proven Zollinger-Ellison syndrome. (Reproduced from ref. 21)

for absolute precision in the radioimmunoassay of peptidal hormones are much less stringent than in the usual clinical laboratory determinations of plasma concentrations of electrolytes, serum proteins or even of steroidal hormones. For instance the normal serum sodium concentrations have a range from 135 to 145 mEq/l; thus the entire normal range has a variation of less than 10% so that an error in the experimental determination of 10% would be intolerable. In the fasting state in normal subjects gastrin concentrations may range from undetectable (<5 pg/ml) to 100 pg/ml and values greater than a thousand-fold higher have been measured in Z-E patients. Diagnostic studies are frequently made not on the basis of a single hormonal concentration but rather on dynamic changes in response to provocative stimuli and suppressants. Therefore it is seldom necessary to have absolute precision of determination to better than 10% and, under some circumstances, errors of considerably greater magnitude are tolerable. When radioimmunoassay is applied to measurement of substances of biologic interest other than the peptidal hormones greater precision may be required and is usually readily achievable.

Over the past decade radioimmunoassay has had its principal impact on endocrinology. Not only has it made possible increased accuracy in the diagnosis of pathologic states characterized by hormonal excess or deficiency but it has been the major tool used in investigations concerned with the regulation of hormonal secretion, with the interrelationships between hormones and with our understanding of hormonal physiology in general. More recently radioimmunoassay has led to the discovery in plasma and tissue of new forms, presumably precursors or products of the commonly recognized peptidal hormones. It is to some aspects of this problem that the remaining portion of this presentation will be directed.

As discussed earlier, the validity of a radioimmunoassay procedure requires that the immunochemical behavior of standards and unknowns be identical. This condition can be tested by making multiple dilutions of an unknown sample and determining whether the curve of competitive inhibition of binding is superposable on the standard curve used for assay. Failure to meet this condition precludes a truly quantitative measurement. Immunochemical heterogeneity was first demonstrated in 1968 when studies from our laboratory established that there were striking immunochemical differences between human parathyroid hormone (hPTH) in plasma and in glandular

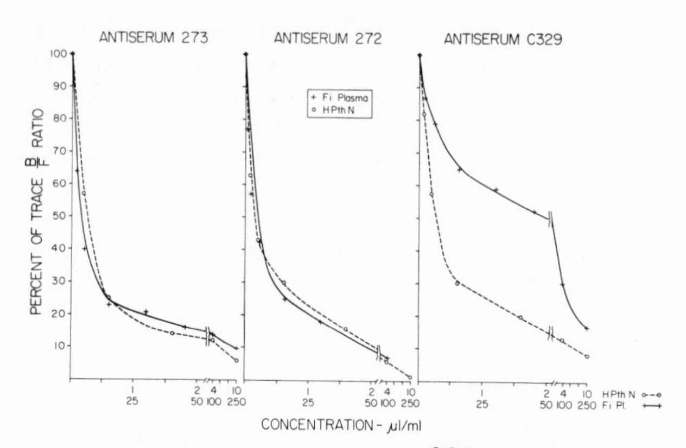

Fig. 11 - Inhibition of binding of ^{125}I-Bpth in three antisera by pooled plasma from a patient with 2° hyperparathyroidism (+) and by extract of a normal parathyroid gland (°). (Reproduced from ref. 22)

extracts (22). Thus while a single factor could be used to superimpose a plasma dilution curve on a curve of standards obtained from a normal parathyroid gland for two antisera, 272 and 273, the same factor resulted in discrepant results with another antiserum, C329 (Fig. 11) (22). Furthermore the rate

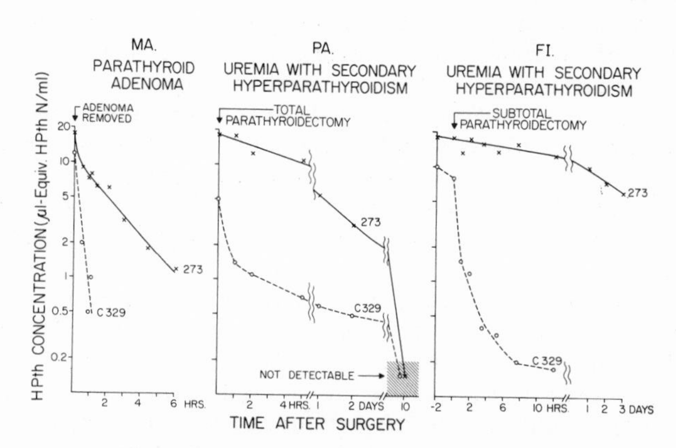

Fig. 12 - Disappearance of immunoreactive parathyroid hormone from plasma following parathyroidectomy in patients with 1° or 2° hyperparathyroidism. Plasma samples were assayed in antiserum C329 and antiserum 273 using extract of a normal human parathyroid gland (hPTH N) as standard and ^{125}I-bPTH as tracer. (Reproduced from ref. 22)

of disappearance of immunoreactivity following parathyroidectomy appeared to depend on the antiserum used for assay (Fig. 12) (22). The heterogeneity of plasma PTH has been widely confirmed by other workers (23-27) as well as in our own laboratory (28). From our studies (28) we have reached the following conclusions: One form of immunoreactive PTH secreted by the gland (Fraction A) (Fig. 13) has Sephadex gel filtration characteristics corresponding to those of intact PTH, bears a reciprocal relationship to serum calcium and disappears from the plasma, even of uremic subjects, with

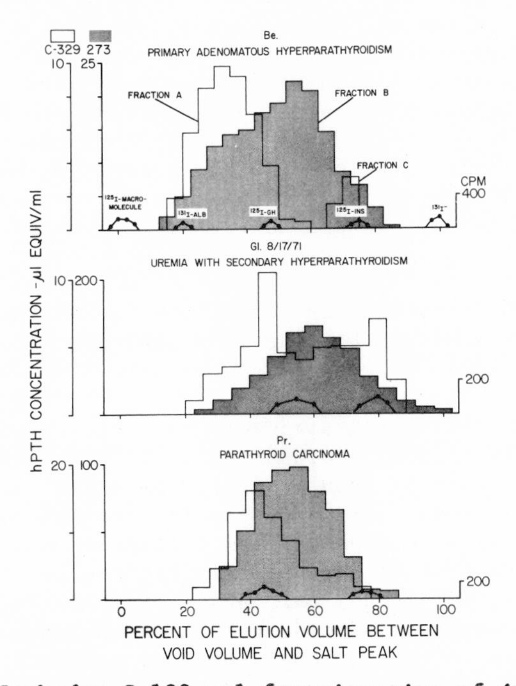

Fig. 13 - Sephadex G-100 gel fractionation of immunoreactive hPTH in plasma of patients with primary adenomatous, secondary uremic, and carcinomatous hyperparathyroidism. Volume of plasma applied to columns was 1 ml in primary and carcinomatous, and 3 ml in secondary hyperparathyroidism. Note variation in magnitude of ordinate scales of antisera 273 and C329 in these fractionations. The component of immunoreactivity that peaked with an elution volume less than that of ^{125}I-hGH was designated Fraction A; that peaked with an elution volume between ^{125}I-hGH and ^{125}I-insulin was designated Fraction B; that peaked with an elution volume equal to or greater than ^{125}I-insulin as Fraction C. (Reproduced from ref. 28)

a half-time of 20 minutes or less. This component reacts strongly with both antisera, 273 and C329, which we usually use and corresponds to the biologically active form of the hormone in the circulation. The predominant component in the plasma, Fraction B, (Fig. 13) is seen primarily by 273, has a molecular weight about 2/3 that of the intact hormone, is probably biologically inactive since it remains elevated even in the presence of clinical postparathyroidectomy hypoparathyroidism and has a disappearance rate from the plasma of a uremic subject more than 100 times longer than that of Fraction A. Fraction B appears to be a C-terminal fragment of the intact hormone. There is also present in plasma another generally minor immunoreactive form (Fraction C) (Fig. 13) which is presumably a N terminal fragment. This component, like B, is biologically inactive and disappears even more slowly than B. We have concluded (28) that the evidence favors a glandular origin for the fragments as well as for the intact hormone and that the relative prominence of the fragments in the plasma arises from their prolonged turnover times.

Dr. Arnaud reviews his recent studies on the heterogeneity of parathyroid hormone in a subsequent paper in this symposium.

After the brilliant discovery by Steiner and associates (29, 30) of pancreatic proinsulin, a biologically relatively inactive precursor of insulin in the pancreas, Roth et al (31) noted in plasma an immunoreactive insulin component ("big" insulin) with Sephadex gel filtration characteristics identical with that of proinsulin. Proinsulin is a single chain peptide, (MW $\sim$9000) half again as large as insulin, in which a connecting peptide runs from the amino terminal of the A chain to the carboxyl terminal of the B chain. It has an elution volume on Sephadex G-50 gel filtration about half that of insulin (Fig. 14). The findings of Roth et al (31) were soon confirmed by Steiner's group (32) and by ourselves (33, 34).

Studies in a number of laboratories have now shown that proinsulin usually comprises only a minor component of total immunoreactive plasma insulin in the stimulated state (Fig. 14). However in some though not in all cases of insulinoma, the major component of immunoreactivity has an elution volume on Sephadex gel corresponding to that of proinsulin (34, 35). Whether the component is intact proinsulin or an intermediate

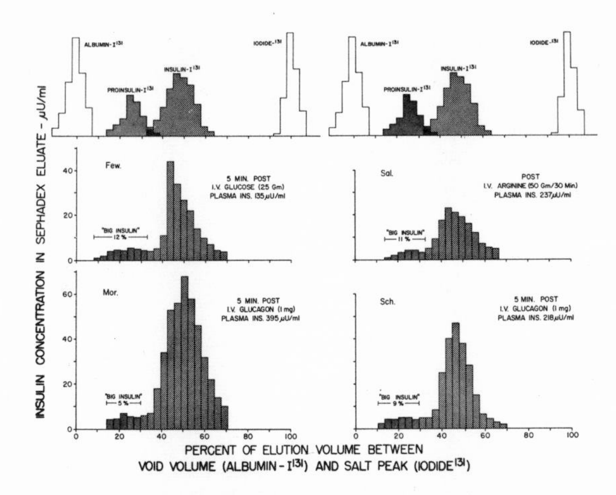

Fig. 14 - Insulin concentration in Sephadex eluates following intravenous glucose, glucagon or arginine stimulation of insulin secretion. The regions of elution of labeled standard insulin and proinsulin are depicted at the top of the figure. (Reproduced from ref. 34)

molecule with an opening at one of the ends of the connecting peptide that joins the A and B chains has not been determined with certainty. Since insulinoma patients usually present with hypoglycemia and the immunoreactive insulin levels are not always markedly elevated, the question arises as to whether the proinsulin-like component could in fact be biologically inactive proinsulin.

We have also demonstrated that a still larger form of insulin, "big, big insulin", predominated in the plasma of an insulinoma suspect (36). This patient was quite unusual in that she had only occasional hypoglycemia in spite of inordinately high plasma immunoreactive insulin (600 μU/ml fasting; 2000 μU/ml post feeding). Thus this new hormonal form of insulin must be devoid of biologic activity in vivo. On Sephadex G-200 gel filtration it has an elution volume smaller than that of labeled albumin and almost coincident with that of labeled γ-globulin (Fig. 15). This new hormonal form of insulin maintains its integrity on refractionation and cannot be distinguished immunochemically from 6000 MW insulin with the antiserum we use for radioimmunoassay. Big big insulin is stable in 8M urea but is rapidly converted by trypsin to an insulin-like component (36). It is a very minor component of extracts of normal pancreas and of insulinomas

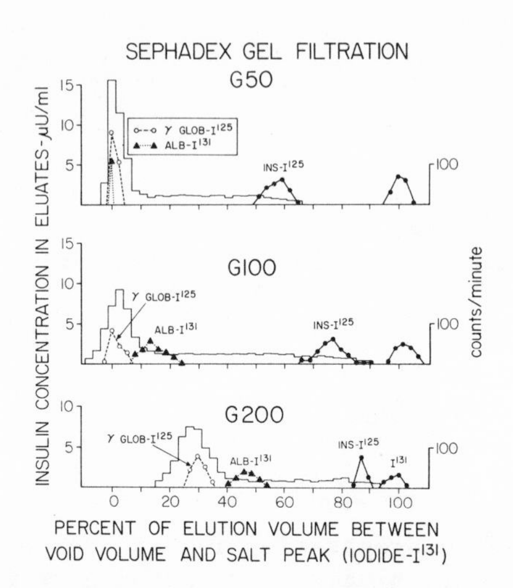

Fig. 15 - Immunoreactive insulin concentrations in Sephadex G-50 (top), G-100 (middle) and G-200 (bottom) eluates of plasma of patient Kc. (Reproduced from ref. 36)

(<1%). The evidence is suggestive that this new hormonal form is a precursor of the insulin family preceding even proinsulin but biosynthetic studies are required to confirm this suggestion.

The nature of the antral hormone, gastrin, was first elucidated only a decade ago by Gregory and Tracy (37) by the extraction and purification from hog antra of two heptadecapeptides, gastrin I and gastrin II. These differ, in the same species, only in the presence of an esterified SO_3H group on the tyrosine in position 12 in gastrin II. In a Conference on Gastrin held in September 1964, these authors had the foresight to state (38) "We have termed the peptides we isolated "Gastrins" I and II, but we do not mean to imply by this that either is considered to be in the same form as the hormone is when released from antral mucosa. Clearly, there may be present in antral mucosa other "gastrins" composed of part of the peptides we have isolated, or indeed incorporating them, or the active parts, within a larger molecule. This consideration must apply also to the substance produced by Zollinger-Ellison tumors." Our studies using the radioimmunoassay of gastrin have confirmed this prediction. When we fractionated the plasma of gastrin hyper-

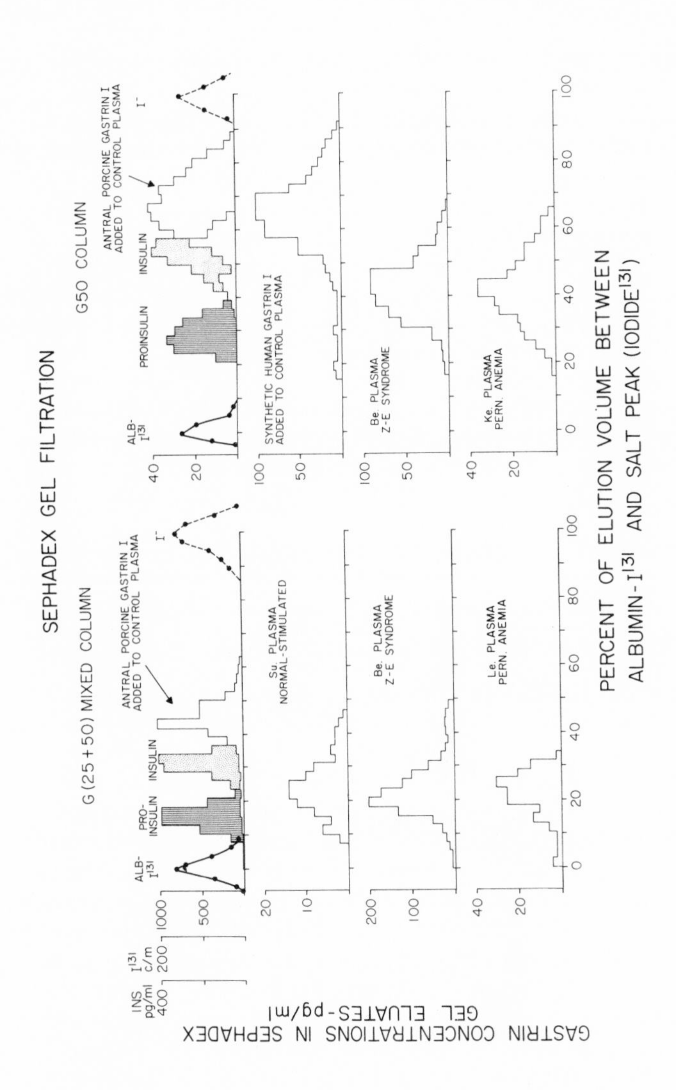

Fig. 16 - Distribution of immunoreactive gastrin in samples of endogenous plasma or plasma-gastrin mixtures added to columns of Sephadex G-50 (right), or mixtures of G-50 and G-25 (left) for gel filtration. The zones of elution of the marker molecules are shown in the top frames. (Reproduced from ref. 39)

secretors (primarily patients with Zollinger-Ellison (Z-E) syndrome or pernicious anemia (PA)), we found that the major component of immunoreactive gastrin in these samples had an elution volume on Sephadex gel filtration between insulin and proinsulin (Fig. 16) and on electrophoretic analysis (Fig. 17) appeared to be more basic than heptadecapeptide gastrin (HG) (39, 40). We termed this new hormonal form "big gastrin" (BG). We demonstrated that with the various antisera we use for radioimmunoassay it was immunochemically indistinguishable from HG; it cannot be transformed to HG by incubation in 8M urea, 2N HCl, or neuraminidase (500 U/ml) but can be converted to HG virtually instantaneously and quantitatively by tryptic digestion (40). Both BG and HG are stimulated by feeding PA patients, although BG disappears from plasma at a slower rate (40). Both BG and HG are present in extracts of antrum and proximal small bowel (41). BG becomes more and more prominent in relation to HG as one proceeds distally down the gastrointestinal tract (Fig. 17) (41). Gregory and Tracy (42) have since purified BG from Z-E tumors, have determined its amino acid composition, and have confirmed the properties which had initially been

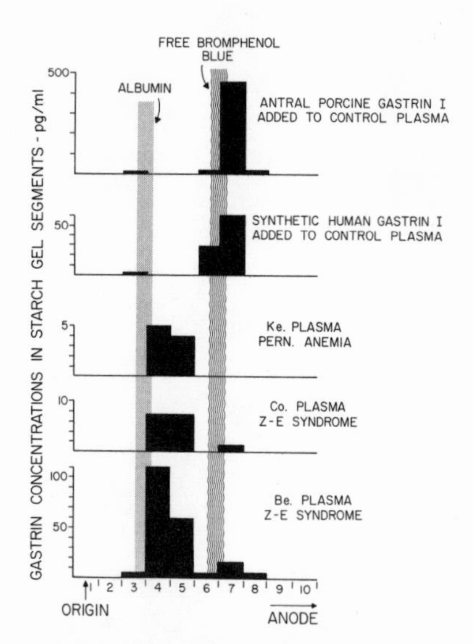

Fig. 17 - Distribution of immunoreactive gastrin components, on starch gel electrophoresis. The zones of migration of bromphenol blue-stained albumin and of free bromphenol blue were noted prior to sectioning. (Reproduced from ref. 39)

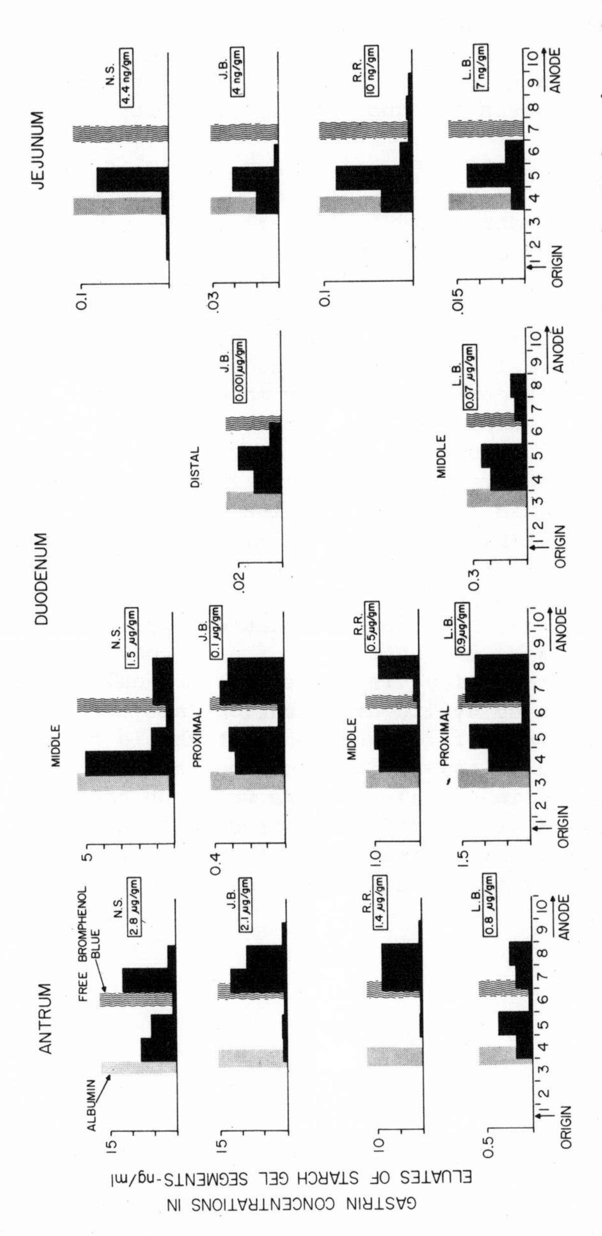

Fig. 18 - Distribution of immunoreactive gastrin. on starch gel electrophoresis, in extracts of antrum, duodenum, and proximal jejunum in postmortem material. The total concentration of gastrin (boxed values) in the crude extract for each sample is expressed as micrograms or nanograms gastrin per g of mucosa. Since gel eluates from different gels were assayed at different dilutions only the relative abundance of the components in each gel is significant. (Reproduced from ref. 41)

determined by radioimmunoassay of picogram to nanogram amounts of immunoreactive gastrin in plasma or tissue extracts containing a millionfold excess of other proteins. Unlike proinsulin, which has low biologic potency, the biologic activity in vivo of BG is about equivalent to that of HG for infusion doses of equal immunoreactivity (43).

Stimulated by our discovery of big big insulin (36) we looked for and found an even larger form of immunoreactive gastrin, "big big" gastrin (BBG) (44, 45). BBG eluted near ^{131}I-albumin in the void volume on Sephadex G-50 gel filtration. It is a minor component (<2% of immunoreactivity) in the plasma of Z-E and PA patients and in extracts of Z-E tumor (44, 45). It is virtually undetectable (<<1% of immunoreactivity) in antral and proximal duodenal extracts. However like BG, it becomes relatively more prominent distally down the gastrointestinal tract, amounting to as much as 24% of total immunoreactivity in some jejunal extracts (44). BBG is a major fraction of immunoreactive gastrin in normal human (Fig. 19), canine and porcine plasma in the non-stimulated states (45). Unlike BG and HG, BBG is not detectably stimulated by feeding (Fig. 19) (45). The half-times for disappearances of HG, BG and BBG from plasma are approximately 3, 9 and 90 minutes respectively (46). Since the plasma concentration of each component under steady state conditions is determined not only by its secretion rate but also by its distribution into extravascular spaces and its rate of degradation, it should be appreciated that even the relatively high concentration of BBG in the fasting state in normal subjects would be consistent with a low secretory rate for BBG compared to BG and HG. Whether BBG is the ultimate precursor of the gastrin family and whether or not it has significant biologic potency have not been evaluated.

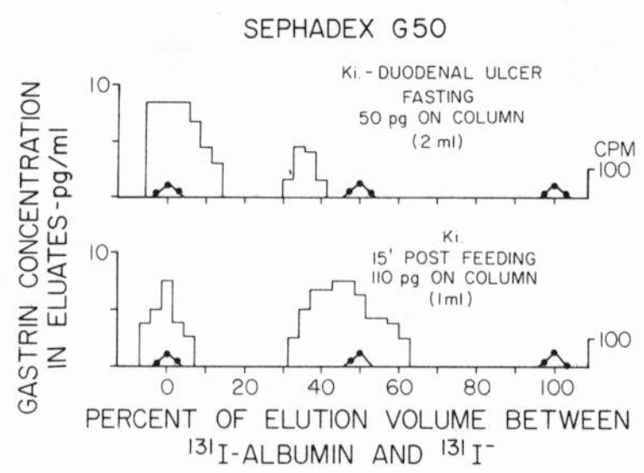

Fig. 19 - Distribution of immunoreactive gastrin on Sephadex G-50 gel filtration in plasma of a human subject before and after feeding. (Reproduced from ref. 45)

Immunoreactive ACTH is also heterogeneous in plasma and tissue (47-50). Highly purified human ACTH (1-39 peptide, "little" ACTH) added to plasma and fractionated on Sephadex G50 columns emerges as a single peak midway between the void volume and the salt peak. However fractionation of plasma or tissue extracts on these columns reveals an immunoreactive component which elutes in or immediately after the void volume. We have designated this component "big" ACTH. In plasma there is great variation in the relative distribution of big and little ACTH (Fig. 20). The only form detectable when plasma ACTH is elevated in response to stimulation of a normal pituitary is little ACTH. This form is found in patients with low plasma cortisol levels whether arising from hypoadrenal function (Addison's disease),

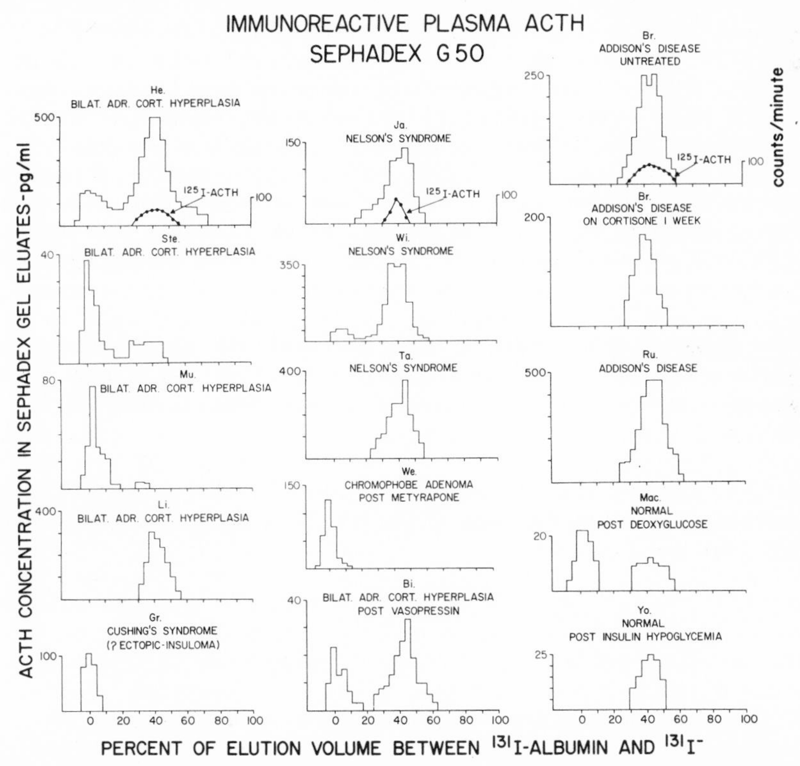

Fig. 20 - Distribution of ACTH immunoreactivity following Sephadex G-50 gel filtration of plasmas. Concentrations of immunoreactive ACTH are shown in open columns. The elution volumes of labeled marker molecules added to the plasma before application of the columns are shown in solid circles. (Reproduced from ref. 48)

(Fig. 20) post bilateral adrenalectomy, or subsequent to administration of metyrapone, an agent that blocks 11 β-hydroxylase activity in the adrenal cortex and leads to a fall in plasma cortisol levels (47, 48, 50). In the plasma of patients with ectopic ACTH production with (47, 48) or without (50) clinical Cushing's syndrome big ACTH predominates. In patients with bilateral adrenal hyperplasia or Nelson's syndrome who are presumed to have autonomous pituitary hypersecretion, the fraction in the big form may range from 0 to 100% (Fig. 20) (48). We have not as yet evaluated what factors are responsible for this very variable fraction in this group of patients.

Big ACTH maintains its integrity on refractionation, is immunochemically indistinguishable from little ACTH with the antiserum we use for radioimmunoassay and, from its behavior on starch gel electrophoresis, appears to be a more acidic peptide than little ACTH (48). Big ACTH is virtually devoid of biologic activity (49) as measured by the adrenal cell dispersion method of Sayers et al (51). Controlled tryptic digestion of big ACTH results almost instantaneously in virtually quantitative conversion to a peptide with many physical chemical characteristics resembling the authentic 1-39 peptide (48) and with biologic activity equivalent to its immunologic activity (49) (Fig. 21).

Although biosynthetic studies have not as yet been performed, these observations are consistent with big ACTH having a precursor relationship to the usual 1-39 ACTH peptide.

A recent observation of considerable interest was the finding of immunoreactive ACTH, predominantly in the big form, in all but one of 30 extracts of primary or metastatic carcinoma of the lung (50). The tissues were obtained as surgical or autopsy specimens from patients with no clinical evidence of Cushing's syndrome. In these same tissues we did not find evidence for ectopic production of human growth hormone, parathyroid hormone, insulin or gastrin. Control extracts of normal lung and lung tissue remote from the tumor did not contain ACTH so that accidental contamination or another non-specific artifact was not responsible (50). Subsequent extractions of another 20 tumor specimens confirm these studies (unpublished observations).

More than half of 83 patients with carcinoma of the

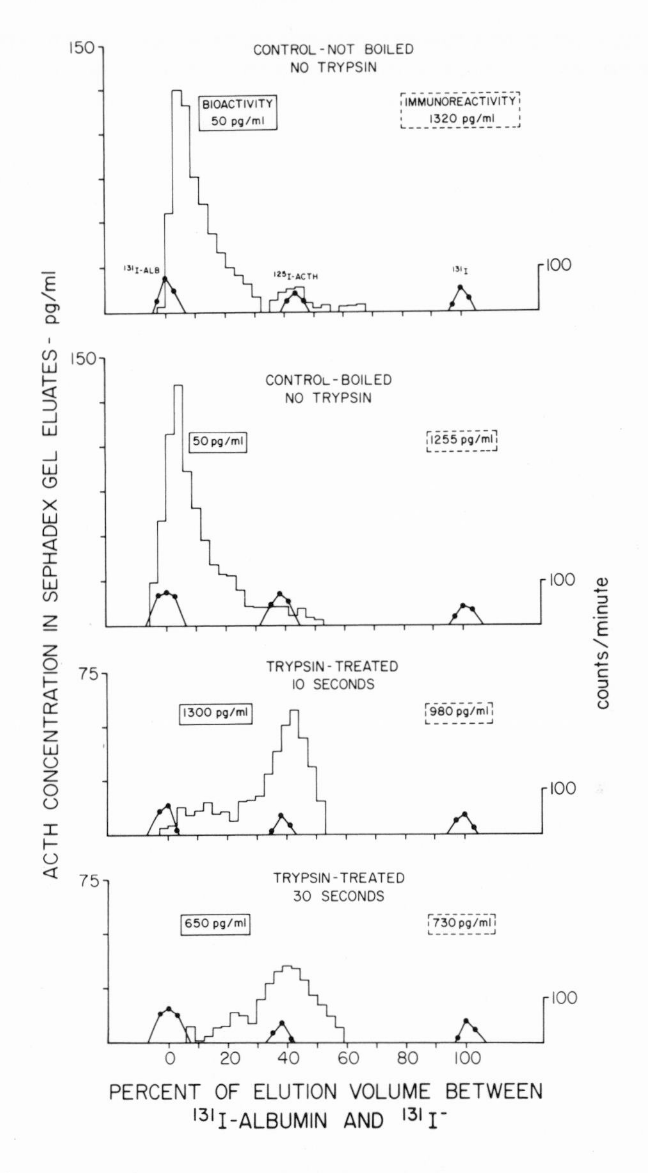

Fig. 21 - Sephadex G-50 gel filtrations of big ACTH (obtained from a boiled water extract of a bronchogenic carcinoma) before and after trypsin treatment. Bioactivity of each sample placed on the columns is indicated in the boxes on the left [solid box] and immunoreactivity in the boxes on the right [dashed box]. (Reproduced from ref. 49)

lung had afternoon plasma ACTH levels greater than 150 pg/ml; more than 90% of those patients with plasma concentrations less than 150 pg/ml had received radiation therapy or chemotherapy for treatment of the tumor (Fig. 22) (50). Only 7% of laboratory controls and other hospital patients without lung disease had plasma ACTH concentrations greater than 150 pg/ml.

Currently under investigation in our laboratory is the possible usefulness of measurement of plasma ACTH concentrations in screening procedures for detection of carcinoma of the lung or as a simple technique for determining the extent and activity of lung tumors and for evaluating the response of such tumors to therapeutic procedures.

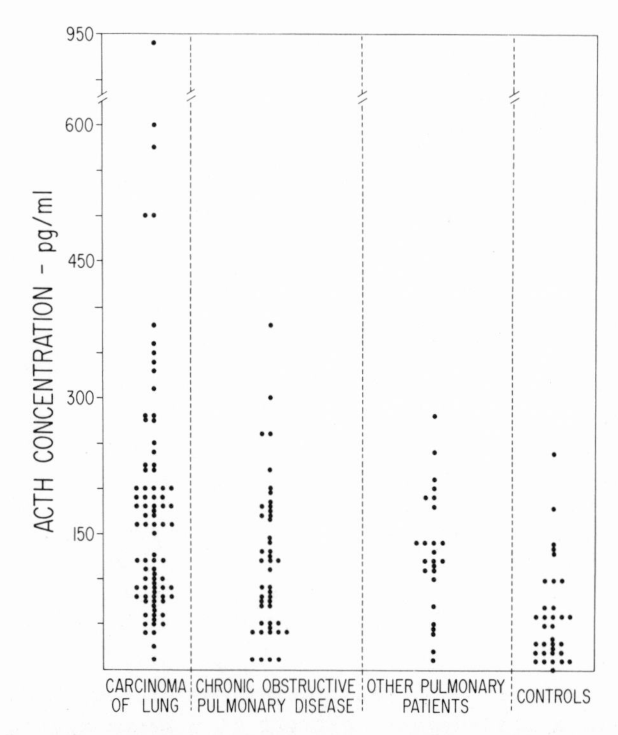

Fig. 22 - Scattergram of immunoreactive ACTH concentrations in afternoon plasma specimens from patients with carcinoma of the lung, chronic obstructive pulmonary disease, other pulmonary patients and controls. (Reproduced from ref. 50)

In this presentation I have selected from past and present endocrinologic investigations of our laboratory in which radioimmunoassay was a necessary tool. These represent but a minute fraction of the total contribution of radioimmunoassay to endocrinology. In other fields the impact of radioimmunoassay is just starting to be felt. It can be anticipated that in the decade to come its burgeoning applications in enzymology, oncology, pharmacology and toxicology and other areas will transcend in importance even its significance in endocrinology. From present indications it seems likely that if there is no other simple way to measure or to detect a substance of biologic interest some imaginative investigator will exploit radioimmunoassay to find a solution to the problem.

REFERENCES

1. Berson, S. A., Yalow, R. S., Post, J., Wisham, L. H., Newerly, K. N., Villazon, M. J. and Vazquez, O.N.: Distribution and fate of intravenously administered modified human globin and its effect on blood volume. Studies utilizing I^{131} tagged globin. J. Clin. Invest. 32:22, 1953.

2. Berson, S. A., Yalow, R. S., Schreiber, S. S. and Post, J.: Tracer experiments with I^{131} labeled human serum albumin: Distribution and degradation studies. J. Clin. Invest. 32:746, 1953.

3. Berson, S. A. and Yalow, R. S.: The distribution of I^{131}-labeled human serum albumin introduced into ascitic fluid: Analysis of the kinetics of a three compartment catenary transfer system in man and speculations on possible sites of degradation. J. Clin. Invest. 33:377, 1954.

4. Mirsky, I. A.: The etiology of diabetes mellitus in man. Recent Progr. Horm. Res. 7:437, 1952.

5. Berson, S. A., Yalow, R. S., Bauman, A., Rothschild, M. A. and Newerly, K.: Insulin-I^{131} metabolism in human subjects: Demonstration of insulin binding globulin in the circulation of insulin-treated subjects. J. Clin. Invest. 35:170, 1956.

6. Topley, W.W.C. and Wilson, G. S., The Principles of Bacteriology and Immunity, ed. 2, Williams and Wilkins Co., Baltimore, 1941, p. 137.

7. Berson, S. A. and Yalow, R. S.: Kinetics of reaction between insulin and insulin-binding antibody. J. Clin. Invest. 36:873, 1957.

8. Berson, S. A. and Berson, S. A.: Quantitative aspects of reaction between insulin and insulin-binding antibody. J. Clin. Invest. 38:1996, 1959.

9. Berson, S. A. and Yalow, R. S.: Species-specificity of human anti-beef pork insulin serum. J. Clin. Invest. 38:2017, 1959.

10. Berson, S. A. and Yalow, R. S.: Isotopic tracers in the study of diabetes. In: Advances in Biol. and Med. Physics, Vol. 6: Acad. Press, N. Y. 1958, p. 349.

11. Yalow, R. S. and Berson, S. A.: Assay of plasma insulin in human subjects by immunological methods. Nature 184:1648, 1959.

12. Yalow, R. S. and Berson, S. A.: Immunoassay of endogenous plasma insulin in man. J. Clin. Invest. 39: 1157, 1960.

13. Berson, S. A. and Yalow, R. S.: Recent studies on insulin-binding antibodies. N.Y. Acad. Sc. 82:338, 1959.

14. Yalow, R. S. and Berson, S. A.: Immunologic specificity of human insulin: Application to immunoassay of insulin. J. Clin. Invest. 40:2190, 1961.

15. Rothenberg, S. P.: Assay of serum vitamin B_{12} concentration using Co^{57}-B_{12} and intrinsic factor. Proc. Soc. Exp. Biol. Med. 108:45, 1961.

16. Barakat, R. S. and Ekins, R. P.: Assay of vitamin B_{12} in blood: A simple method. Lancet 2:25, 1961.

17. Ekins, R. P.: The estimation of thyroxine in human plasma by an electrophoretic technique. Clin. Chim. Acta 5:453, 1960.

18. Murphy, B.E. P. Application of the property of protein-binding to the assay of minute quantities of hormones and other substances. Nature 201, 679, 1964.

19. Roth, J.: Peptide hormone binding to receptors: A review of direct studies in vitro. Metabolism 22: 1059, 1973.

20. Yalow, R. S. and Berson, S. A.: Radioimmunoassay of gastrin. Gastroenterology 58:1, 1970.

21. Berson, S. A. and Yalow, R. S.: Radioimmunoassay in gastroenterology. Gastroenterology 62:1061, 1972.

22. Berson, S. A. and Yalow, R. S.: Immunochemical heterogeneity of parathyroid hormone in plasma. J. Clin. Endocrinol. Metab. 28:1037, 1968.

23. Arnaud, C. D., Sizemore, G. W., Oldham, S. B., Fisher, J. A., Tsao, H. S. and Littledike, E. T.: Human parathyroid hormone: Glandular and secreted molecular species. Amer. J. Med. 50:630, 1971.

24. Arnaud, C. D., Tsao, H. S. and Oldham, S. B.: Native human parathyroid hormone: An immunochemical investigation. Proc. Nat. Acad. Sci. 67:415, 1970.

25. Canterbury, J. M. and Reiss, E.: Multiple immunoreactive molecular forms of parathyroid hormone in human serum. Proc. Soc. Exp. Biol. Med. 140:1393, 1972.

26. Habener, J. F., Powell, D., Murray, T. M., Mayer, G. P. and Potts, J. T., Jr.: Parathyroid hormone: secretion and metabolism in vivo. Proc. Nat. Acad. Sci. 68:2986, 1971.

27. Sherwood, L. M., Lundberg, W. B., Jr., Targovnik, J. H. Rodman, J. S. and Seyfer, A.: Synthesis and secretion of parathyroid hormone in vitro. Amer. J. Med. 50:658, 1971.

28. Silverman, R. and Yalow, R. S.: Heterogeneity of parathyroid hormone: Clinical and physiologic implications. J. Clin. Invest. 52:1958, 1973.

29. Steiner, D. F. and Oyer, P. E.: The biosynthesis of insulin and a probable precursor of insulin by human islet cell adenoma. Proc. Nat. Acad. Sci. 57:473, 1967.

30. Steiner, D. F., Cunningham, D., Spigelman, L. and Aten, B.: Insulin biosynthesis: Evidence for a precursor. Science 157:697, 1967.

31. Roth, J., Gorden, P. and Pastan, I.: "Big insulin": A new component of plasma insulin detected by immunoassay. Proc. Nat. Acad. Sci. 61:138, 1968.

32. Rubenstein, A. H., Cho, S. and Steiner, D. F.: Evidence for proinsulin in human urine and serum. Lancet 1: 1353, 1968.

33. Yalow, R. S. and Berson, S. A.: Dynamics of insulin in early diabetes - in humans. In: Early Diabetes, Supplement 1 to Advances in Metabolic Disorders, Proc. of the First International Symposium (Oct. 23-26, 1968) R. A. Camerini - Davalos and H. S. Cole, eds., Acad. Press, N. Y. 1970, p. 95.

34. Goldsmith, S. J., Yalow, R. S. and Berson, S. A.: Significance of human plasma insulin Sephadex fractions. Diabetes 18:834, 1969.

35. Yalow, R. S. and Berson, S. A.: Fundamental principles of radioimmunoassay techniques in measurement of hormones. In: Recent Advances in Endocrinology, Proc. 7th Pan-American Congress of Endocrinology (Aug. 16-21, 1970) E. Mattar, G. B. Mattar and V. H.T. James, eds., Excerpta Medica, ICS 238, 1971, p. 16.

36. Yalow, R. S. and Berson, S. A.: "Big, big insulin". Metabolism 22:703, 1973.

37. Gregory, R. A. and Tracy, H. J.: The constitution and properties of two gastrins extracted from hog and antral mucosa. I. The isolation of two gastrins from hog and antral mucosa. Gut 5:103, 1964.

38. Gregory, R. A. and Tracy, H. J.: Studies on the chemistry of gastrins I and II. In: Gastrin. UCLA Forum in Medical Sciences, No. 5. M. I. Grossman, ed., Univ. of Calif. Press, Berkeley and Los Angeles, 1966, p. 9.

39. Yalow, R. S. and Berson, S. A.: Size and charge distinctions between endogenous human plasma gastrin in peripheral blood and heptadecapeptide gastrins. Gastroenterology 58:609, 1970.

40. Yalow, R. S. and Berson, S. A.: Further studies on the nature of immunoreactive gastrin in human plasma. Gastroenterology 60:203, 1971.

41. Berson, S. A. and Yalow, R. S.: Nature of immunoreactive gastrin extracted from tissues of gastrointestinal tract. Gastroenterology 60:215,1971.

42. Gregory, R. A. and Tracy, H. J.: Isolation of two "big gastrins" from Zollinger-Ellison Tumour tissue. Lancet 2:797, 1972.

43. Walsh, J. H., Debas, H. T. and Grossman, M. I.: Pure natural human big gastrin: Biological activity and half life in dog. Gastroenterology 64:A-187/873, 1973.

44. Yalow, R. S. and Berson, S. A.: And now, "big, big" gastrin. Biochem. Biophys. Res. Commun. 48:391, 1972.

45. Yalow, R. S. and Wu, N.: Additional studies on the nature of big big gastrin. Gastroenterology 65:19, 1973.

46. Straus, E. and Yalow, R. S.: Studies on the distribution and degradation of heptadecapeptide, big, and big big gastrins. Gastroenterology 1974 (in press).

47. Yalow, R. S. and Berson, S. A.: Size heterogeneity of immunoreactive human ACTH in plasma and in extracts of pituitary glands and ACTH-producing thymoma. Biochem. Biophys. Res. Commun. 44:439, 1971.

48. Yalow, R. S. and Berson, S. A.: Characteristics of "big ACTH" in human plasma and pituitary extracts. J. Clin. Endocrinol. Metab. 36:415, 1973.

49. Gewirtz, G., Schneider, B., Krieger, D. T. and Yalow, R. S.: Big ACTH: Conversion to biologically active ACTH by trypsin. J. Clin. Endocrinol. Metab. 1974 (in press).

50. Gewirtz, G. and Yalow, R. S.: Ectopic ACTH production in carcinoma of the lung. J. Clin. Invest. 1974 (in press).

51. Sayers, G., Swallow, R. L. and Giordano, N. D.: An improved technique for the preparation of isolated rat adrenal cells: A sensitive, accurate and specific method for the assay of ACTH. Endocrinology 88:1063, 1971.

MEASUREMENT OF BIOGENIC AMINES AT THE PICOGRAM LEVEL

Juan M. Saavedra

Visiting Scientist
National Institutes of Health
Bethesda, Maryland 20014

SUMMARY

A number of enzymatic-isotopic methods for the assay of biogenic amines at the picogram level have been recently described. These assays allow the measurement of dopamine, norepinephrine, histamine, β-phenylethylamine, octopamine, phenylethanolamine, serotonin, N-acetylserotonin, and tryptamine in small areas of the brain and in body fluids. They are based on the incubation of the amines with specific methyltransferase enzymes, together with a donor of radioactive methyl groups, 3H-methyl-S-adenosyl-l-methionine. The enzymatically formed radioactive N or O methyl products are separated by solvent extraction and the radioactivity is counted. These assays are useful to determine the concentrations of putative neurotransmitters in individual brain nuclei of the rat brain, in isolated neurons from molluscs, and to study low levels of biogenic amines in human plasma under different physiological and pathological conditions.

INTRODUCTION

Over the last years, our laboratory has been involved in the development of sensitive enzymatic methods to measure biogenic amines in tissues and small amounts of body fluids, and in the application of these methods to the study of the metabolism of these amines under physiological and pathological conditions. These methods are based on the incubation of the biological material with specific methyltransferase enzymes (1,2,3,4,5,6,7,8,11,12,13,14,15,16,18) together with the donor of methyl groups, S-adenosyl-l-methionine. The enzymes catalyse the transfer of the radioactive methyl group of S-adenosyl-l-methionine to the O or N terminal positions of the amines, resulting in a formation of O or N methyl derivatives. The derivatives are then separated by means of selective solvent extraction procedures, and the radioactivity is counted.

GENERAL PROCEDURE

The general procedure for the measurement of biogenic amines is outlined in Table 1.

TABLE 1

Procedure for the Enzymatic-Isotopic Assay for Biogenic Amines

1. Extraction of the amine from the tissue or body fluid (acid or buffer extraction)
2. Incubation with corresponding N-methyltransferase and radioactive methyl donor
3. Extraction of radioactive N- or O-methylated product formed in the reaction with an organic solvent
4. Elimination of radioactive contaminants by selective drying procedures
5. Counting of the radioactivity by liquid scintillation spectroscopy

The two major considerations in setting up an assay for biogenic amines are specificity of the assay and sensitivity. Specificity of the assays was obtained by a) the use of methylating enzymes with specificity for a given amine or group of amines, b) the use of solvents of different degrees of polarity to separate the radioactive product formed in the reaction, c) the application of evaporation techniques to eliminate radioactive but volatile contaminating substances. The use of 3H-methyl-S-adenosyl-l-methionine with high specific activity has recently provided additional degree of sensitivity to assays already available (Table 2).

TABLE 2

Sensitivity of the Enzymatic-Isotopic Methods for the Determination of Biogenic Amines in Tissues

Amine	Enzyme	Sensitivity (picograms)
Tryptamine	NMT	1000
N-Acetylserotonin	HIOMT	50
Serotonin	NAT - HIOMT	50
Phenylethanolamine	PNMT	25
Octopamine	PNMT	25
β-Phenylethylamine	DBH - PNMT	200
Histamine	HMT	25
Dopamine	COMT	100
Norepinephrine	COMT	10

NMT = Nonspecific N-methyltransferase from rabbit lung
HIOMT = Hydroxyindole O-methyl transferase (EC 2.1.1.4) from bovine pineal gland
NAT = N-acetyltransferase (EC 2.3.1.5) from rat liver
PNMT = Phenylethanolamine N-methyltransferase from bovine adrenal
DBH = Dopamine-β-hydroxylase from bovine adrenal
HMT = Histamine N-methyltransferase (EC 2.1.1.c) from guinea pig brain
COMT = Catechol-O-methyltransferase (EC 2.1.1.a) from rat liver

ENZYME PREPARATION

The methyltransferase enzymes are partially purified by ammonium sulphate fractionation and dialysis (Table 3). By the use of simple purification procedures, relatively stable enzyme preparations are obtained. Most methyltransferases are stable at this stage of purification for at least one month, when stored at $-20^{o}C$. The use of highly purified enzymes is generally not essential to ensure specificity in these assays. In the case of dopamine-β-hydroxylase, the application of sephadex chromatography is required to eliminate endogenous inhibitors of the reaction (14).

TABLE 3

Enzymes Used in the Detection of Biogenic Amines

Enzyme	Source	Purification Procedure	Amine Measured
Histamine-N-methyl-transferase (EC 3.1.1.c)	guinea pig brain	1-2	Histamine
Catechol-O-methyl-transferase (EC 2.1.1.d)	rat liver	1-2	Dopamine Noradrenaline
Dopamine-β-hydroxylase (DBH) (EC 1.14.2.1)	bovine adrenal	1-2-3	
Nonspecific N-methyl-transferase (NMT)	rabbit lung	1-2	Tryptamine
Hydroxyindole O-methyl-transferase (EC 2.1.1.4) (HIOMT)	bovine pineal	1-2	N-acetyl-serotonin; Serotonin
N-acetyltransferase (EC 2.3.1.5) (NAT)	rat liver	1-2	Serotonin
Phenylethanolamine-N-methyltransferase (PNMT)	bovine adrenal	1-2	Phenylethanol-amine; Octop-amine; Pheny-lethylamine

Purification Procedures: 1. Ammonium sulphate fractionation
2. Dialysis
3. Sephadex chromatography

EXTRACTION OF THE AMINES FROM TISSUES

Tissue samples are homogenized in several volumes of buffer, alkaline or acid solutions, according to the amine. After homogenization, the amines are measured in aliquots of the supernatant fluid. In the case of β-phenylethylamine, a preliminary extraction procedure is necessary (14). Internal standards are carried through the entire procedure, and the results are corrected for recoveries.

INCUBATION AND EXTRACTION PROCEDURES

Aliquots of the tissue supernatant fractions are incubated with partially purified methyltransferases (Table 3), in the presence of ^{3}H-methyl S-adenosyl-l-methionine. After the enzymatic reaction is completed, the ^{3}H-methyl derivatives formed are separated from the radioactive methyl donor by means of solvent extraction procedures. The use of relatively non polar solvents allow a rapid and in most cases complete separation of the methylated amines from S-adenosyl-l-methionine. A proportionality between the product formation and the amount of amine originally present in the tissue is always established (Fig. 1).

USE OF DOUBLE ENZYMATIC TECHNIQUES

In the case of serotonin and β-phenylethylamine, the direct methylation cannot be conveniently performed. The amines are first enzymatically converted into derivatives with higher affinity for methyltransferase enzymes, and the methylation is then performed (Fig. 2 and 3).

The use of double enzymatic techniques confers a high degree of specificity to the methods, and allows the use of tissue blanks, by the omission of essential co-factors in the preliminary step (13,14).

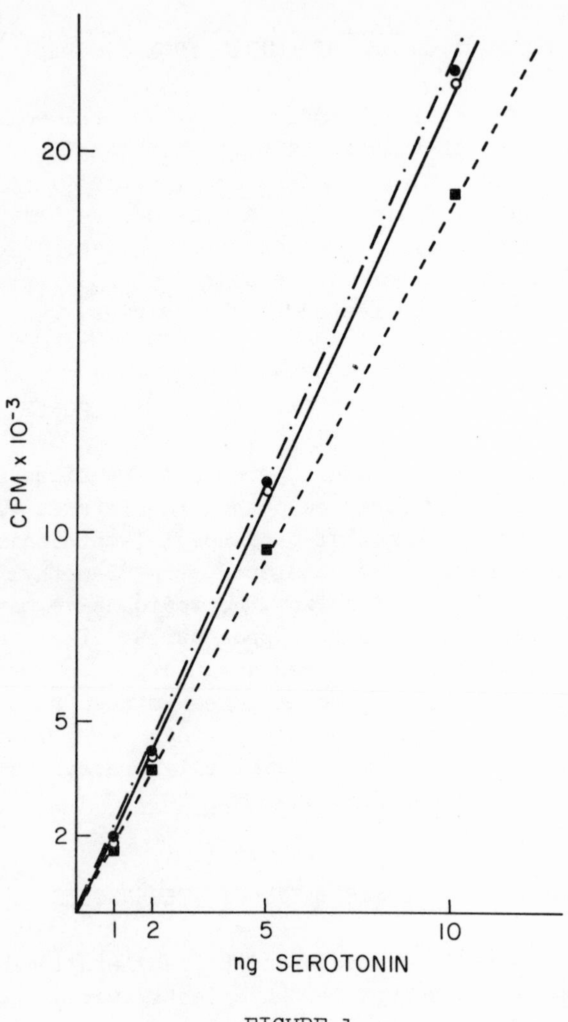

FIGURE 1

Linearity of Serotonin Recoveries from Tissues

Organs were homogenized in 0.01N HCl, centrifuged and divided into aliquots. Various amounts of serotonin were added and carried through the procedure. Results are expressed in counts per minute after substraction of the blank and serotonin content. ● aqueous media; ■ rat brain ○ rat pineal gland.

FIGURE 2

Enzymatic-Isotopic Assay for Serotonin

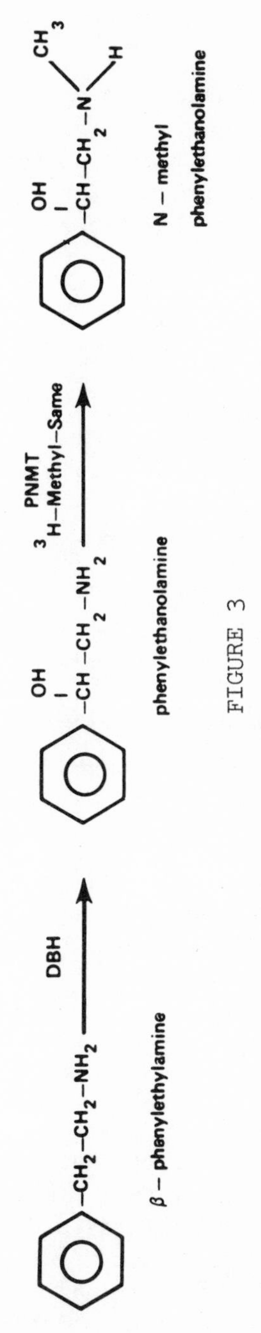

FIGURE 3

Enzymatic-Isotopic Assay for β-Phenylethylamine

CONDITIONS FOR SPECIFICITY

a) Identification of the Product Formed in the Reaction

The specificity of the assay is examined by identification of the product formed in the enzymatic reaction, by means of thin layer chromatography in several solvent systems (Fig. 4).

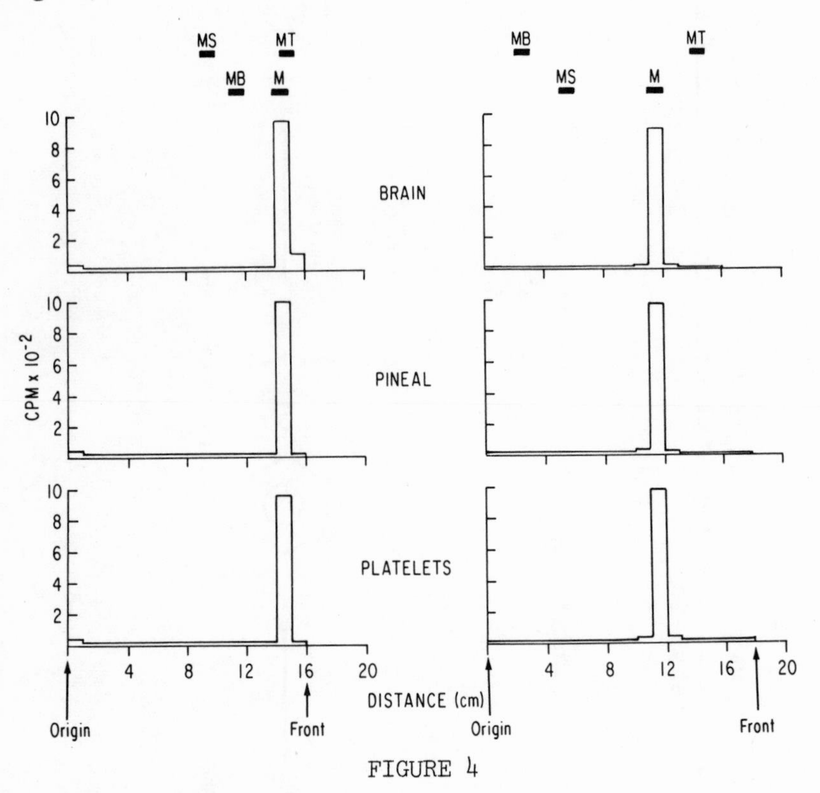

FIGURE 4

Specificity of the Serotonin Assay

Thin layer chromatography of the product obtained from rat brain and pineal gland, and human platelets. Solvent used: methylacetate-isopropanol-ammonium hydroxide 10% (45:35:20) (left in the figure); toluene-acetic acid-ethyl acetate-water (80:40:21:5) (right in figure). Standards used were melatonin (M), O-methyltryptophol (MT) O-methyltryptamine (MS) and O-methyl N, N dimethyltryptamine (MB).

b) Interference of Related Compounds

The possible interference of related compounds is checked by the determination of the degree of interference when these compounds are carried through the assay.

Table 4 represents the specificity of the serotonin assay. Endogenous N-acetylserotonin cannot be differentiated from N-acetylserotonin formed in the reaction (Fig. 2). The _in vitro_ formed and the endogenous N-acetylserotonin can be differentiated by eliminating acetyl co-enzyme A.

TABLE 4

Specificity of the Serotonin Enzymatic-Isotopic Assay

Compound	Amount ng	Radioactivity cpm	% of serotonin value (1 ng)
Serotonin	1	2,200	
Serotonin	10	20,150	
N-acetylserotonin	1	3,280	149
5-hydroxytryptophol	1	0	0
Bufotenin	1	0	0
5-hydroxyindole acetic acid	10	0	0
5-hydroxytryptophan	1	1,050	50
5-hydroxytryptophan +MK 486 (4×10^{-5}M)	1	0	0

Tissue 5-hydroxytryptophan can be decarboxylated to serotonin by the L-amino acid decarboxylase present in the liver giving a false high value of serotonin (17). The addition of the decarboxylase inhibitor MK-486 (4×10^{-5}M) prevents 5-hydroxytryptophan from interfering with the serotonin assay (Table 4).

APPLICATION OF THE ASSAYS TO PHARMACOLOGICAL AND CLINICAL STUDIES

The enzymatic-isotopic assays have been applied to a number of animal and clinical studies. It was possible to detect and measure for the first time octopamine, β-phenylethylamine, phenylethanolamine and tryptamine in the brain (11,12,14).

The recent development of a technique for the dissection of individual brain nuclei (9), allowed the quantitation of dopamine, norepinephrine, histamine and serotonin in 16 different nuclei from the rat hypothalamus (7,10,17). Table 5 shows the distribution of biogenic amines in the rat median eminence. These results suggest that the biogenic amines studied could be related to the regulation of neuroendocrine function.

TABLE 5

Biogenic Amines in the Rat Median Eminence

	Concentration* ng/mg protein
Serotonin	15.3 ± 3.2
Histamine	17.8 ± 2.2
Norepinephrine	29.5 ± 4.0
Dopamine	65.0 ± 6.1

*Data from Brownstein _et al_. (7), Palkovits _et al_. (10) and Saavedra _et al_. (17).

Some of the methods are currently applied to the study of biogenic amines in humans β-phenylethylamine and tryptamine have been recently detected in human plasma, in small amounts (1 to 5 ng/ml) (Saavedra and Axelrod, unpublished observations).

CONCLUSIONS

Specific and sensitive enzymatic-isotopic methods for the measurement of biogenic amines have been recently developed. The enzymatic-isotopic assays have several advantages: They are fairly simple to perform, relatively inexpensive, do not require sophisticated equipment, and they are more sensitive than most of the other assays for biogenic amines available today. About 50 to 100 assays can be done by one person in a day.

The assays can be widely used in pharmacology and neurobiology. The levels and metabolism of biogenic amines can be studied in small areas of the brain, such as individual nuclei in the rat hypothalamus and limbic system. The possible presence of putative neurotransmitters can be studied in isolated neurons of molluscs (Brownstein, Saavedra, Carpenter, and Axelrod, in preparation).

Some of the assays can be applied to clinical studies in man. Low levels of several biogenic amines can be detected in fluids. Studies on the changes in amine levels in the human plasma, under different physiological and pathological conditioning are currently in progress.

REFERENCES

1. AXELROD, J. and TOMCHICK, R. Enzymatic O-methylation of epinephrine and other catechols. J. Biol. Chem., 233, 702-705 (1958).
2. AXELROD, J. and WEISSBACH, H. Purification and properties of hydroxyindole O-methyl transferase. J. Biol. Chem. 236, 211-213 (1961).
3. AXELROD, J. Purification and properties of phenylethanolamine N-methyl transferase. J. Biol. Chem. 237, 1657-1660 (1962).
4. AXELROD, J. The enzymatic N-methylation of serotonin and other amines. J. Pharmac. Exp. Ther. 138, 28-33 (1962).

5. BROWN, D. D., TOMCHICK, R., and AXELROD, J. Distribution and properties of a histamine-methylating enzyme. J. Biol. Chem. 234, 2948-2950 (1959).
6. BROWNSTEIN, M., SAAVEDRA, J. M., and AXELROD, J. Control of N-acetylserotonin by a β-adrenergic receptor. Mol. Pharmac. 9, 605-611 (1973).
7. BROWNSTEIN, M., SAAVEDRA, J. M., PALKOVITS, M., and AXELROD, J. Histamine content of hypothalamic nuclei of the rat. Submitted for publication.
8. COYLE, J. T. and HENRY, D. Catecholamines in fetal and newborn rat brain. J. Neurochem. 21, 61-67 (1973).
9. PALKOVITS, M. Isolated removal of hypothalamic or other brain nuclei of the rat. Brain Research 59, 449-450 (1973).
10. PALKOVITS, M., BROWNSTEIN, M., SAAVEDRA, J.M., and AXELROD, J. Norepinephrine and dopamine content of hypothalamic nuclei. Submitted for publication.
11. SAAVEDRA, J. M. and AXELROD, J. A specific and sensitive enzymatic assay for tryptamine in tissues. J. Pharmac. Exp. Ther. 183, 363-369 (1972).
12. SAAVEDRA, J. M. and AXELROD, J. Demonstration and distribution of phenylethanolamine in brain and other tissues. Proc. Natn. Acad. Sci., U.S.A., 70, 769-772 (1973).
13. SAAVEDRA, J. M.,BROWNSTEIN, M., and AXELROD, J. A specific and sensitive enzymatic-isotopic microassay for serotonin in tissues. J. Pharm. Exp. Ther. 186, 508-515 (1973).
14. SAAVEDRA, J. M. Enzymatic-isotopic assay for and presence of β-phenylethylamine in brain. J. Neurochem. (in press).
15. SAAVEDRA, J. M. Enzymatic-isotopic method for octopamine at the picogram level. Anal. Biochem. (in press).
16. SAAVEDRA, J. M. and AXELROD, J. Enzymatic-isotopic micromethods for the measurement of biogenic amines in brain tissue and body fluids. J. Psychiat. Research. (in press).
17. SAAVEDRA, J. M., PALKOVITS, M., BROWNSTEIN, M., and AXELROD, J. Serotonin distribution in the nuclei of the rat hypothalamus and preoptic region. Submitted for publication.
18. SYNDER, S. H., BALDESSARINI, R., and AXELROD, J. A sensitive and specific enzymatic isotopic assay for tissue histamine. J. Pharm. Exp. Ther. 153, 544-549 (1966).

PARATHYROID HORMONE: STRUCTURE AND IMMUNOHETEROGENEITY

Claude E. Arnaud, M.D. & H. Bryan Brewer, Jr., M.D.
Mineral Research Laboratory and the Department of Endocrine Research, Mayo Clinic and Mayo Foundation, Rochester, Minnesota 55901, and the National Institutes of Health, Bethesda, Maryland 20014

During the past 5 years there has been a great increase in our understanding of the chemistry, biosynthesis, secretion, and pathophysiology of parathyroid hormone (PTH). PTH has been shown by Hamilton, Cohn, Kemper, and colleagues[1-4] to be synthesized as a precursor with a molecular weight of approximately 12,000. Structural studies of the bovine proparathyroid hormone (pro-PTH) have shown six additional amino acid residues attached to the NH_2-terminal end of the molecule.[5] Because amino acid compositional data suggest that pro-PTH has amino acid residues not yet accounted for by this hexapeptide sequence, the possibility that there is an additional sequence of amino acids attached to the COOH-terminal region is now being considered.[6,7] It is likely that the pro-PTH is converted, in the parathyroid gland, to the storage form of the hormone, an 84 amino acid polypeptide with a molecular weight of 9,500. It is not known presently if pro-PTH is released into the circulation.

The 84 amino acid polypeptide is a major form of the hormone secreted by the gland into the circulation after appropriate physiologic stimuli.[8] However, recent studies by a number of groups[6,9-19] indicate that there are multiple forms of the hormone in the blood of patients with hyperparathyroidism. These multiple forms include the 84 amino acid polypeptide and fragments of it. The major fragment has a molecular weight of 6,000 to 7,000, is COOH-terminal, and has a very long half-life. Canterbury et at.[20] have demonstrated that a fragment(s) in the plasma of hyperparathyroid patients was biologically active in the renal adenylate cyclase system, which suggests that some of the circulating plasma fragments retain biologic activity. Indisputable immunologic evidence of this latter observation is still lacking, although it is supported by work, presented in the present paper, using an antiserum of great specificity for NH_2-terminal PTH.

The present paper describes recent studies on the chemistry of human, bovine, and porcine PTH as well as some practical consequences of the immunoheterogeneity of PTH in serum when this radioimmunoassay is used in the routine evaluation of patients with hyperparathyroidism.

CHEMISTRY OF PARATHYROID HORMONE

Human Parathyroid Hormone

The human PTH (HPTH) used in our studies was isolated from parathyroid adenomas obtained from patients undergoing surgery for hyperparathyroidism. A trichloroacetic acid extract[21] of the human adenomas was purified by gel filtration followed by ion exchange chromatography on CM-Sephadex. The isolated hormone was shown to be homogeneous by disc gel electrophoresis and Edman NH_2-terminal analysis.[22]

The NH_2-terminal sequence analysis of HPTH was performed on a Beckman Sequencer (model 890B) using 1 M Quadrol buffer. The phenylthiohydantoin derivatives of the amino acids obtained from the automated Edman degradations were identified by gas-liquid chromatography[23] and mass spectroscopy.[24, 25] Chemical ionization (CI) mass spectroscopy was performed on a Finnigan quadrupole mass spectrometer equipped with a PDP-8/e digital computer and a Complot plotter. Electron impact (EI) mass spectrometry was performed on an LKB mass spectrometer (model 9000).

The sequence of the initial 34 amino acids of HPTH is shown in Figure 1. A detailed description of the sequence analysis of HPTH including the CI mass spectra of each of the individual steps in the sequence has been published.[22]

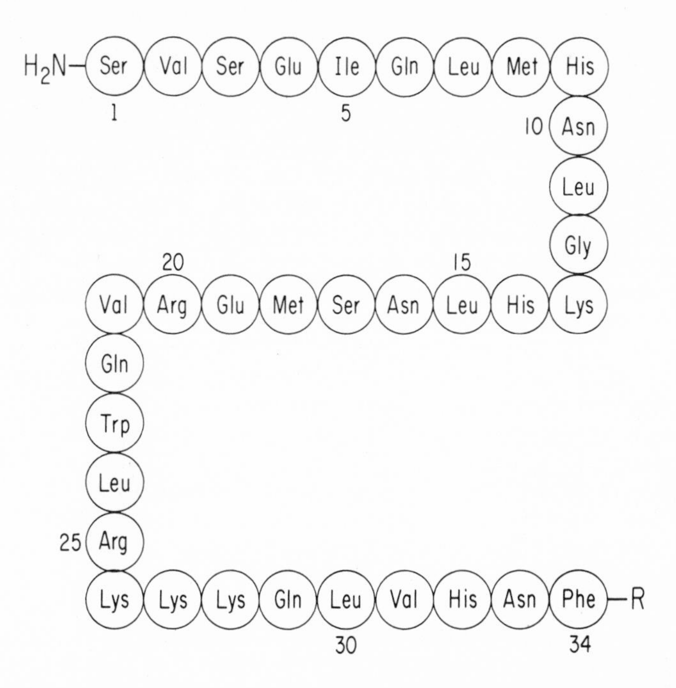

Figure 1. Initial 34 amino acid residues of human parathyroid hormone. (From Brewer HB Jr, Fairwell T, Rittel W, Littledike T, Arnaud CD: Recent studies on the chemistry of human, bovine, and porcine parathyroid hormones. Am. J. Med. [in press]. By permission of Dun·Donnelley Publishing Corporation.)

Bovine Parathyroid Hormone

The complete sequence of the 84 amino acid bovine hormone has been reported.[26,27] In those studies we identified residue 22 as glutamic acid. However, the human sequence contained glutamine at position 22, and so we reexamined position 22 in the bovine hormone.[28] The possibility existed that the glutamine was deamidated during the purification or the Edman degradation.

Bovine PTH was isolated from defatted parathyroid tissue with care to avoid acidic conditions for prolonged periods in order to minimize possible deamidation of glutamine. In addition, the temperature and cleavage time for conversion of the thiazolinone to the thiohydantoin were decreased during the Edman degradation at step 22 in the sequence. Three separate automated degradations were performed on the bovine hormone--one on the intact hormone and two on the isolated COOH-terminal cyanogen bromide peptide (residues 19 through 84). The cyanogen bromide peptide was utilized in these studies in order to decrease the number of Edman cycles required to reach step 22 in the sequence (22 cycles for the intact hormone; 4 cylces for the cyanogen bromide peptide) and thereby decrease the number of times that the polypeptide was exposed to heptafluorobutyric acid (HFBA). HFBA is the acid used during the cleavage step in the automated Edman procedure, and repeated exposure to it at elevated temperature (55°C) has been associated with deamidation of glutamine residues. In all three of these automated degradations, glutamine was identified at position 22. Three techniques--CI mass spectroscopy, EI mass spectroscopy, and gas-liquid chromatography--were used to confirm the identification of glutamine at this position (Figure 2).

Porcine Parathyroid Hormone

O'Riordan et al.[29] have previously reported the presence of a glutamic acid residue at position 22 in the sequence of porcine PTH. The presence of glutamine at position 22 in both human and bovine PTH suggested that the porcine hormone also might have glutamine at this position. We therefore reexamined the sequence of porcine PTH, using the same precautions to minimize deamidation that were used for the bovine hormone. However, porcine PTH was difficult to purify to homogeneity, and repetitive chromatography on CM-Sephadex was required to obtaine hormone that was homogeneous by disc gel electrophoresis and NH_2-terminal analysis.

Automated Edman sequence analysis of purified porcine PTH revealed glutamine at position 22. CI mass spectroscopy and gas-liquid chromatography indicated that approximately two-thirds of the glutamine residues at position 22 had undergone deamidation to glutamic acid during purification or sequence analysis. EI mass spectroscopy (Figure 3) revealed only glutamic acid (m/e 264), reflecting the extensive deamidation. These results indicated that the glutamic acid function reported in porcine PTH at position 22 was also glutamine. Furthermore, the glutamine at position

22 appears to be relatively labile and may undergo extensive deamidation during structural analysis.

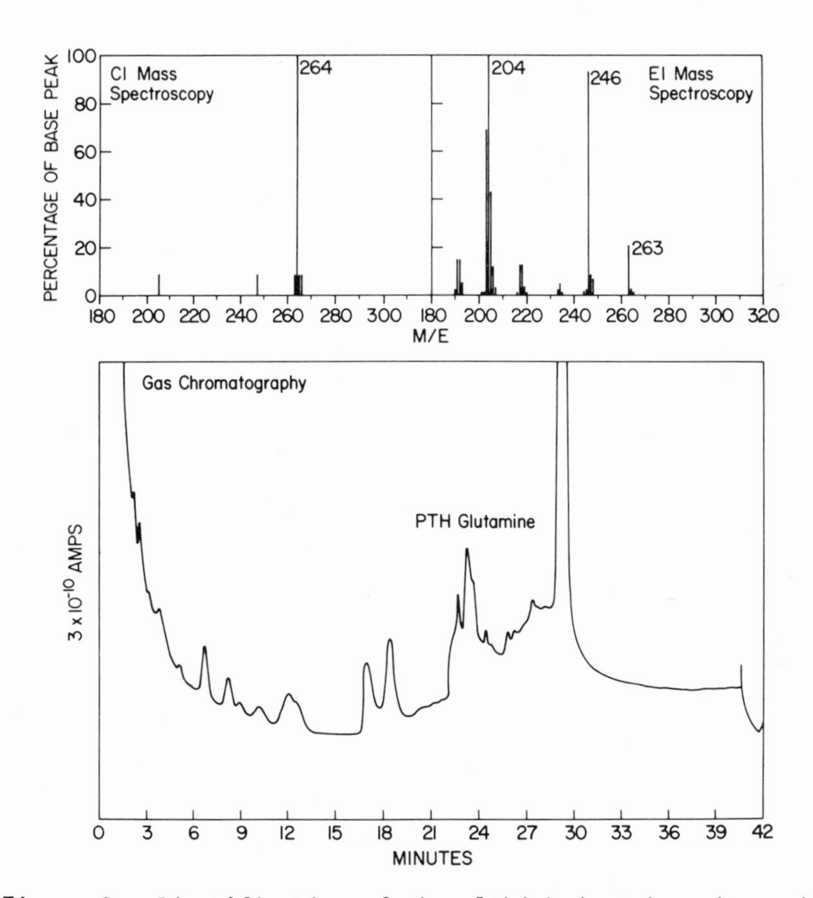

Figure 2. Identification of phenylthiohydantoin amino acid derivative obtained at step 22 in automated Edman degradation of bovine parathyroid hormone. Glutamine was identified in the CI spectrum at mass 264 and in the EI spectrum at mass 263. The fragmentary ions at m/e 204 and 246 in the EI spectrum are derived from the thermal decomposition of glutamine. (From Brewer HB Jr, Fairwell T, Rittel W, Littledike T, Arnaud CD: Recent studies on the chemistry of human, bovine, and porcine parathyroid hormones. Am. J. Med. [in press]. By permission of Dun·Donnelley Publishing Corp.).

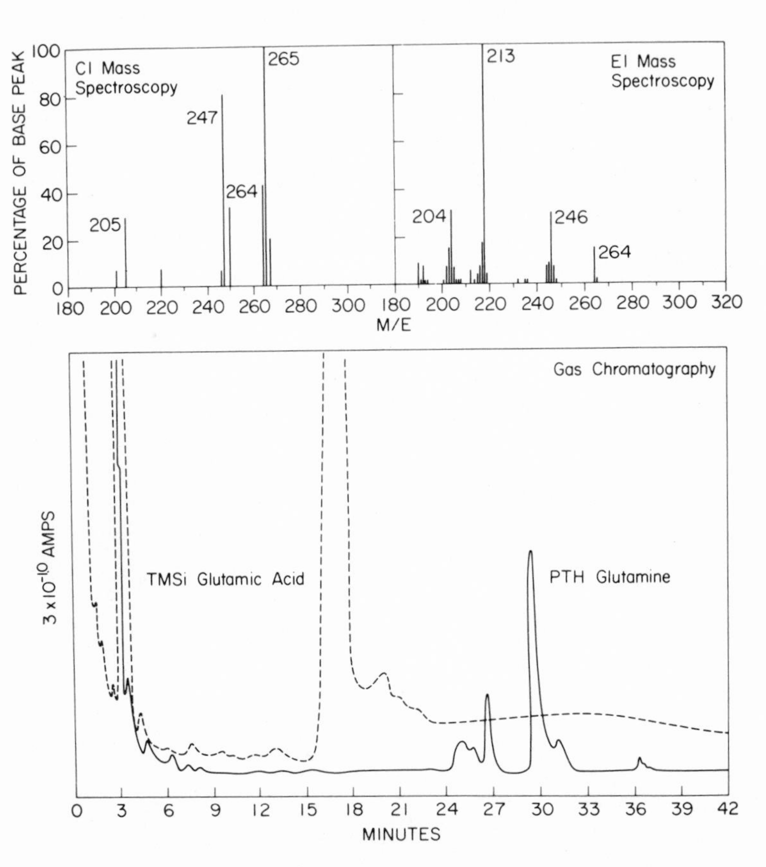

Figure 3. Identification of phenylthiohydantoin amino acid derivatives obtained at step 22 in automated Edman degradation of porcine parathyroid hormone. Glutamine and glutamic acid were identified at m/e 264 and 265, respectively, in the CI spectrum. In the EI spectrum, only glutamic acid (m/e 264) was observed, due to the extensive deamidation of the glutamine residue. Fragmentary ions derived from the thermal decomposition of glutamine are seen at m/e 205 and 247 in the CI spectrum and at m/e 204, 213, and 246 in the EI spectrum. Glutamine and glutamic acid were observed in the gas chromatogram of step 22. In the gas chromatogram,

the solid line represents the profile obtained after silylation of the sample. (From Brewer HB Jr, Fairwell T, Rittel W, Littledike T, Arnaud CD: Recent studies on the chemistry of human, bovine, and procine parathyroid hormones. Am. J. Med. [in press]. By permission of Dun·Donnelley Publishing Corporation.)

The revised sequences of bovine and porcine PTH are shown in Figure 4. The bovine and porcine hormones differ in only 7 of the 84 positions: residues 1, 7, 18, 42, 43, 47, and 74.

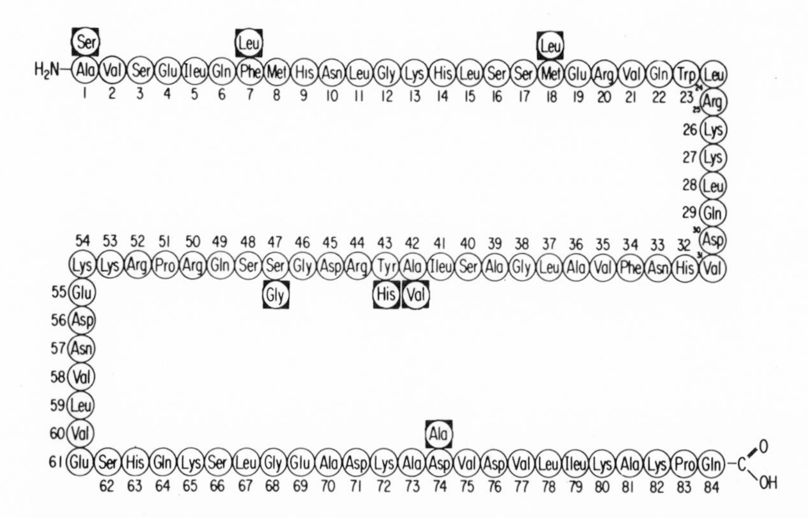

Figure 4. Revised sequences of bovine (circles) and porcine (squares) parathyroid hormone, showing glutamine at position 22. The structure of the bovine and porcine hormones differ by only seven amino acid residues. (From Brewer HB Jr, Fairwell T, Rittel W, Littledike T, Arnaud CD: Recent studies on the chemistry of human, bovine, and porcine parathyroid hormones. Am. J. Med. [in press]. By permission of Dun·Donnelley Publishing Corporation.)

Structural Differences in Human, Bovine, and Porcine PTH 1-34

Previous studies indicated that the intact 84 amino acid polypeptide is not required for biologic activity; the biologic activity appears to be within the first 27 residues of bovine PTH and the first 30 residues of porcine PTH.[30] In the first 34 residues, HPTH differs from bovine PTH by 5 residues and from porcine PTH by 4 residues (Figure 5). The initial 15 residues of human and porcine PTH are identical; bovine PTH differs by an alanine residue at position 1 and a phenylalanine residue at position 7. The human and bovine hormones contain two methionine residues, whereas the porcine hormone contains a single methionine, at position 8. All three hormones have a glutamine at position 22. The amino acid residues of the NH_2-terminal region that are unique to the human sequence are an asparagine at position 16, a lysine at position 28, and a leucine at position 30.

HUMAN PARATHYROID HORMONE

BOVINE PARATHYROID HORMONE

PORCINE PARATHYROID HORMONE

Figure 5. Amino acid sequences of NH_2-terminal 34 residues of human, bovine, and porcine parathyroid hormones. (From Brewer HB Jr, Fairwell T, Rittel W, Littledike T, Arnaud CD: Recent studies on the chemistry of human, bovine, and porcine parathyroid hormones. Am. J. Med. [in press]. By permission of Dun·Donnelley Publishing Corporation.)

Potts et al.[31] recently described a sequence for the initial 37 residues of HPTH that differs in three positions from our sequence. The differences include glutamic acid rather than glutamine at position 22, leucine rather than lysine at position 28, and aspartic acid rather than leucine at position 30. In addition, alternate residues, threonine at position 22 and serine at position 32, were identified. The differences between these sequences are as yet unexplained but may reflect isohormonal variation or differences in methods.

Biologic Activity of Synthetic Human PTH 1-34

Andreatta et al.[32] synthesized, by classical techniques, a biologically active peptide containing the initial 34 residues of HPTH, based on the sequence shown in Figure 1. When perfused into the conscious thyroparathyroidectomized rat, this synthetic peptide produced hypercalcemia (6 to 11 mg/dl), hyperphosphaturia (0.3 to 2.4 mg P/h), and hypocalciuria (0.3 to 0.1 mg Ca/h) (Figure 6). However, on in vitro assay of the synthetic peptides in human renal cortical membranes, bovine PTH 1-34 stimulated adenylate cyclase 10 to 30 times greater than did HPTH 1-34 (Figure 7). These studies indicate that use of different assay systems to measure the biologic activity of PTH may give discordant results. Therefore, one must be cautious in interpreting the biologic activity of different hormone preparations based on a single type of assay system.

IMMUNOHETEROGENEITY OF PTH IN HUMAN CIRCULATION

History and Statement of Problem

In 1968, Berson and Yalow[9] published an extremely important paper in which they presented strong evidence in support of the immunoheterogeneity of PTH in human plasma. They showed that one of their anti-bovine-PTH sera distinguished between the hormone extracted from human parathyroid tumors and that present in the peripheral plasma of patients with hyperparathyroidism and that values for the concentration of immunoreactive PTH (iPTH) in plasma differed markedly depending on the antiserum used in the radioimmunoassays. Further support for their suggestion that these phenomena

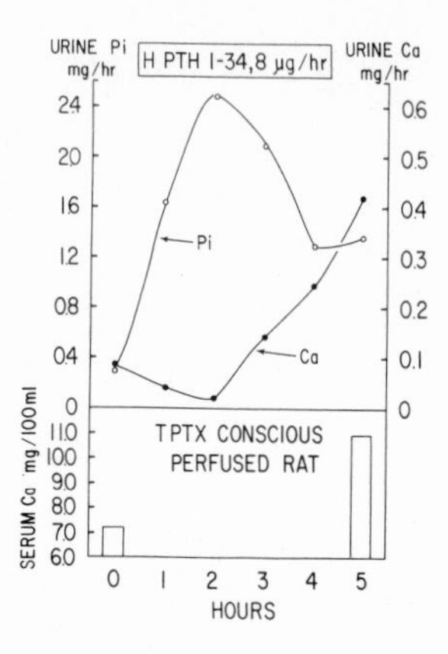

Figure 6. Typical responses of urinary phosphorus (o-o), urinary calcium (●-●), and serum calcium (bar graph, lower panel) to intravenous infusion of synthetic human PTH 1-34 (Brewer et al. sequence) in a thyroparathyroidectomized rat. (From Brewer HB Jr, Fairwell T, Rittel W, Littledike T, Arnaud CD: Recent studies on the chemistry of human, bovine, and porcine parathyroid hormones. Am J Med [in press]. By permission of Dun·Donnelley Publishing Corporation.)

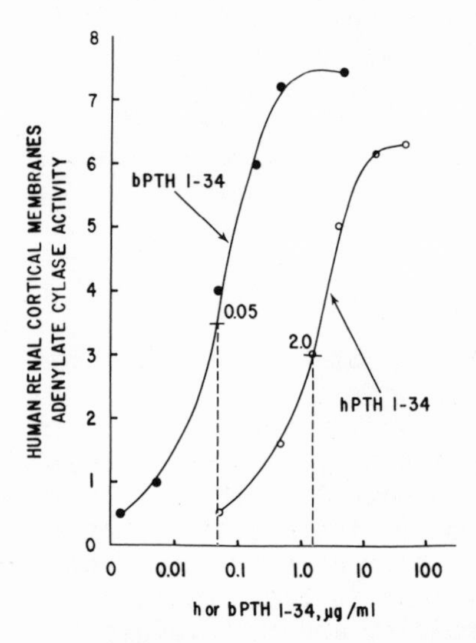

Figure 7. Comparative dose-response curves of synthetic bovine PTH 1-34 (glutamic acid residue at position 22 (●-●) and human PTH 1-34 (Brewer et al. sequence) (o-o) in human renal cortical adenylate cyclase assay system.

reflected immunoheterogeneity of plasma PTH was obtained by them with their observation that the measured half-life of disappearance of iPTH from plasma after parathyroidectomy in a patient with secondary hyperparathyroidism and chronic renal failure was different when different antisera were used in the assays of iPTH. They interpreted their data as indicating the presence of metabolites of PTH rather than isohormones of PTH in the peripheral circulation of hyperparathyroid man.

At about the time that their paper appeared, Reiss and Canterbury[33] reported the development of a radioimmunoassay for PTH in human serum which was capable of measuring iPTH in all normal subjects and of distinguishing clearly between normals and patients with adenomatous hyperparathyroidism. This latter observation was in sharp contrast with that of Berson and Yalow[34] who found a large overlap in plasma iPTH values between these two groups. It is likely that this apparent discrepancy was the first clue that the phenomenon of immunoheterogeneity of plasma PTH has great practical significance in the application of the radioimmunoassay of PTH to problems of clinical diagnosis and investigation.

The second clue came from another discrepancy between the results obtained with different antisera in different laboratories. Reiss and Canterbury[35] reported that it was possible to distinguish between adenoma and primary hyperparathyroidism due to hyperplasia by measurement of serum iPTH during induced hypercalcemia. With this manipulation, hormone concentrations decreased in patients with hyperplasia but did not in patients with adenoma. In contrast, using a radioimmunoassay with another antiserum, Potts and co-workers[36] found that serum iPTH increased with induced hypocalcemia and decreased with induced hypercalcemia in patients with adenomatous hyperparathyroidism.

We do not think that these discrepancies were due to differences in the experimental procedures used in the different laboratories or to the use of reagents derived from nonhuman species (bovine in each instance). As will be shown below, it is likely that the key variable in the explanation is antibody specificity and the differences in the rates of metabolism of the various molecular species of PTH in the circulation.

Initial attempts to characterize and identify the molecular species of PTH present in the circulation of hyperparathyroid man were made by Arnaud and his co-workers[10,11] using cultures of parathyroid tumor explants in vitro. They showed that multiple immunoreactive species of PTH were present in the media of these cultures and ranged in size from molecular weights of 5,000 to >10,000; they suggested at that time that one source of the immunoheterogeneity of serum PTH might be parathyroid tissue itself.

Direct study of this problem was first reported by Canterbury and Reiss.[12] They showed, by gel filtration of Amicon-filtered, pooled, peripheral hyperparathyroid serum, that there are at least three immunoreactive forms of PTH in this serum. One form was eluted from Bio-Rad P-10 columns with ^{131}I-labeled bovine PTH (84 amino acids; mol. wt. 9,500) and had a half-life of less than 30 minutes, whereas the other two forms (approximate mol. wt. 7,000 to 8,000 and 4,000 to 5,000) had half-lives in the range of hours. Habener and associates[8] gel filtered whole serum from hyperparathyroid subjects and confirmed the presence of the 9,500 mol. wt. and the 7,000 to 8,000 mol. wt. forms of the hormone in peripheral serum (but not the 4,000 to 5,000 mol. wt. species of Canterbury and Reiss)[12] and also showed that, in serum obtained from small thyroid veins, the 9,500 mol. wt. form of the hormone predominated.

Recently, Habener and co-workers[13] and Segre and co-workers[14] studied hyperparathyroid sera by using radioimmunoassays based on anti-bovine sera preadsorbed with fragments of bovine PTH and presumably specific for the NH_2-terminal (residues 14 through 19) and COOH-terminal (residues 53 through 84) regions of the bovine PTH molecule. They found that most of the immunoreactivity in these sera was of the COOH-terminal type. Because the synthetic NH_2-terminal fragment (residues 1 through 34) of bovine PTH is biologically active but the COOH-terminal fragment (residues 53 through 84) obtained from natural sources is biologically inert,[37] these workers concluded that the major portion of circulating PTH is biologically inactive. Although this work is interesting and important, it should be recognized that neither study[13,14] detected the 4,000 to 5,000 mol. wt. circulating iPTH fragment previously reported by Canterbury and Reiss[12] and, in a recent paper, Canterbury and associate[20] report that this fragment has biologic activity in the rat renal adenylate cyclase assay system. A model encompassing our intepretation of current knowledge in this field and incorporating information presented below is presented in Figure 8.

Practical Importance in Assessment of Parathyroid Function in Man

We have assessed parathyroid function in patients with hyperparathyroidism by measuring serum iPTH with two radioimmunoassays for PTH--one with high relative sensitivity and specificity for the NH_2-terminal region and one with high relative sensitivity and specificity for the COOH-terminal region of the human PTH molecule--and have correlated these results with data on the number of osteoclasts in bone biopsy specimens from the same patients.[18]

The methods we use in the radioimmunoassay of PTH in human serum have been described.[39] Figure 9 shows the specificity characteristics of the two assay systems with respect to their ability to react with synthetic human and bovine PTH 1-34. The assay system that uses CH 14M antiserum (chicken anti-bovine-PTH serum) reacts almost equally well with the two synthetic peptides. (In studies not shown here, it reacted almost as well with the two synthetic peptides as with the respective native hormones obtained from natural sources.) In contrast, the assay system that uses GP 1M antiserum (guinea pig anti-porcine-PTH serum) does not react with either synthetic human or bovine PTH 1-34 but reacts with human PTH -184 with identical affinity as the assay system that uses CH 14M antiserum. From these and other studies, we have concluded that the assay system using CH 14M antiserum has relatively high specificity for the NH_2-terminal region of human PTH and is identified as "anti N" and that the assay system using GP 1M antiserum has relatively high specificity for the COOH-terminal region of human PTH and is identified as "anti C" in the figures and text of this paper.

The "anti N" and "anti C" assay systems were used to identify the various immunoreactive species of PTH in the effluent fractions from the gel filtration of whole peripheral serum from a patient with primary hyperparathyroidism

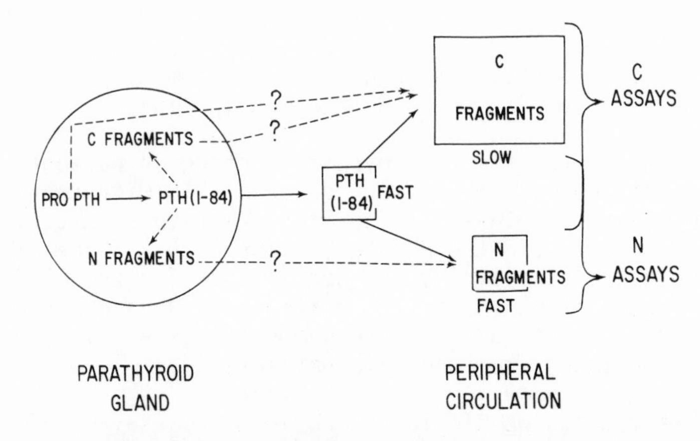

Figure 8. Simplified schema of current concept of the metabolic alteration in PTH (1-84) that result in immunoheterogeneity. Proparathyroid hormone (proPTH) (mol. wt. $\geq$ 11,000) is converted to PTH (1-84) (mol. wt. = 9,500) by specific enzymatic cleavage. It is not presently known if proPTH is released from the gland into the blood. PTH (1-84) is the major secreted species of PTH, but there is evidence that the gland contains a calcium-regulated enzyme that can cleave PTH (1-84) to COOH-terminal (C) and NH_2-terminal (N) fragments.[38] These fragments also might be released into the circulation, but there is no direct evidence for this. Once secreted, PTH (1-84) is converted peripherally into C and N fragments. The C fragments have a slow turnover rate. N-specific radioimmunoassays of serum iPTH give low values and C-specific radioimmunoassays give high values because of the differences in the pool sizes of these species of iPTH. This scheme is based on data obtained from the study of serum from hyperparathyroid patients. Data on normal serum are not available. (From Arnaud CD: Parathyroid hormone: coming of age in clinical medicine. Am. J. Med. 55:577-581, 1973. By permission of Dun·Donnelley Publishing Corporation.)

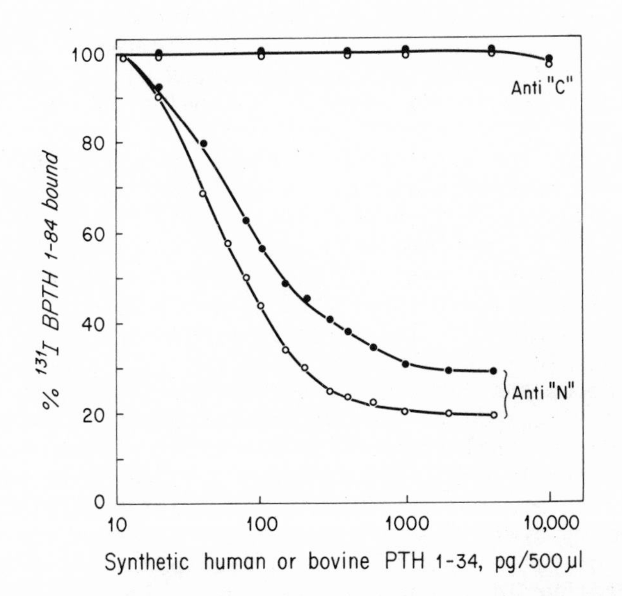

Figure 9. Standard curves demonstrating the relative reactivity of GP 1M (anti-C) antiserum and CH 14M (anti-N) antiserum with synthetic human (•–•) and bovine (o-o) PTH (1-34). (From Arnaud CD, Goldsmith RS, Bordier PJ, Sizemore GW: Influence of immunoheterogeneity of circulating PTH on results of radioimmunoassay of serum parathyroid hormone in man. Am J. Med. [in press]. By permission of Dun·Donnelley Publishing Corporation.)

(Figure 10). As expected from the specificity characteristics of the two assay systems, both reacted equally well with an iPTH component eluting in a position similar to that of ^{131}I-labeled bovine PTH 1-84. However, only the "anti C" assay system reacted with an iPTH component that eluted between ^{131}I-labeled bovine PTH 1-84 and bovine PTH 1-34 and that was present in much greater quantities than the component eluting the position of PTH 1-84. Both the "anti N" and the "anti C" system reacted with a component that eluted immediately after ^{131}I-labeled bovine PTH 1-34.

Although the resolution of these components by these methods is incomplete and the general configurations of their individual elution profiles suggest further heterogeneity within the components, we have concluded that the peripheral serum of patients with primary hyperparathyroidism contains at least three (and probably more) immunoreactive species of PTH. The first component probably represents endogenous PTH 1-84, although it is possible that it is heterogeneous and contains some secreted biosynthetic precursor of PTH as well. The second component is probably a large COOH-terminal fragment of PTH 1-84. This component has consistently comprised at least 5 to 20 times the quantity of iPTH eluting with PTH 1-84 in sera from patients with primary or secondary hyperparathyroidism; however, in the sera of patients with ectopic hyperparathyroidism due to nonparathyroid cancer, there is only 1/10 to 1/7 as much of this component as in sera from patients with primary hyperparathyroidism.[40] The third component appears to be recognized by both antisera. But, if these antisera have absolute "anti C" and "anti N" specificities for the human PTH molecule, it is necessary to conclude that this component is heterogeneous and consists of at least one COOH-terminal and one NH_2-terminal fragment that come off the column together. If, on the other hand, the antisera are not absolutely specific and have overlapping recognition sites, this third component might represent a single, small COOH-terminal fragment. We favor the former alternative.

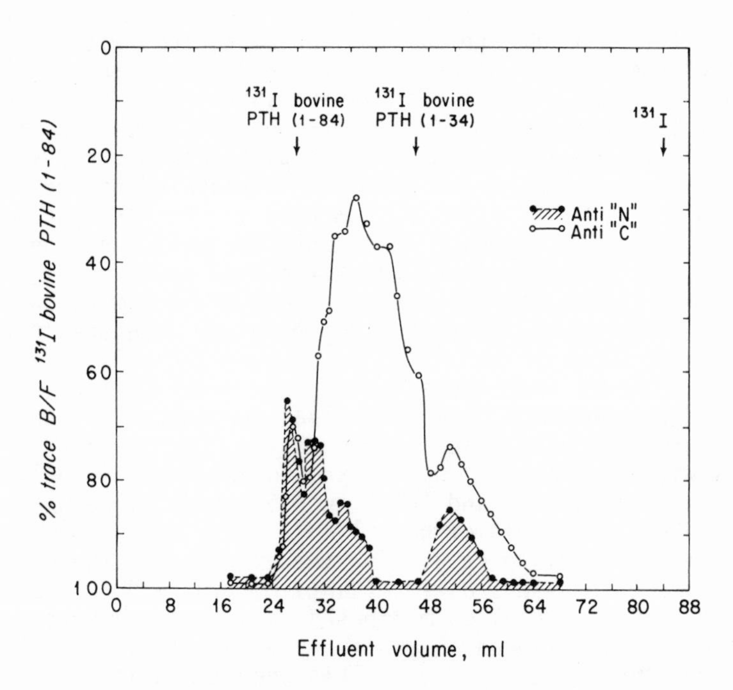

Figure 10. Elution profile of iPTH assayed with GP 1M (anti-C) (o-o) and CH 14M (anti-N) (• • and shaded area) after gel filtration of standard hyperparathyroid plasma (B1) on Bio-Gel P-30. Arrows indicate elution positions of ^{131}I-labeled bovine PTH (1-84), ^{131}I-labeled bovine PTH (1-34), and ^{131}I-. (From Arnaud CD, Goldsmith RS, Bordier PJ, Sizemore GW: Influence of immunoheterogeneity of circulating PTH on results of radioimmunoassay of serum parathyroid hormone in man. Am. J. Med. [in press]. By permission of Dun·Donnelley Publishing Corporation.)

The results of measurements of serum iPTH in hyperparathyroid and normal subjects with the "anti C" and "anti N" specific radioimmunoassays are shown in Figures 11, 12 and 13. The standard curves in Figure 11 compare the reactivities of the "anti C" and "anti N" assay systems with highly purified human PTH 1-84 and the PTH in hyperparathyroid serum. It is clear that both assay systems have the same reactivities with human PTH 1-84 (Figure 11 left). However, only 1/10 to 1/5 the amount of hyperparathyroid serum is required to inhibit the binding of ^{131}I-labeled bovine PTH 1-84 in the "anti C" assay system than in the "anti N" assay system (Figure 11 right). This confirms, with whole serum, the observation made in the gel filtration studies (Figure 9) that in hyperparathyroid serum there is much more COOH-terminal immunoreactivity than NH_2-terminal immunoreactivity.

In patients with end-stage chronic renal failure, the "anti C" assay system gave PTH values ranging from 3 to 200 times normal in all patients (Figure 12). In contrast, assays with the "anti N" system showed increases in only 42% of the same patients, and these increases ranged between barely above the limit of normal and 3 to 4 times normal. Although the absolute values for the extremes of the normal ranges for both assays differ ("anti C," 10 to 50 μl eq/ml; "anti N," 50 to 280 μl eq/ml; based on a standard hyperparathyroid plasma assigned an arbitrary value of 1,000 μl eq/ml), their relative ranges do not. This phenomenon is related to the large quantity of COOH-terminal relative to NH_2-terminal immunoreactivity in the hyperparathyroid serum used as standard in both assay systems.

When serum iPTH values in normal subjects and in patients with primary hyperparathyroidism were plotted as a function of the total calcium concentration measured in the same serum sample (Figure 13), a negative relationship (r = -0.568; P <0.001) was observed over the normal serum calcium range with the "anti C" assay system. The relationship was not significant with the "anti N" assay system (r = -0.171; P <0.1). Serum iPTH values were in the normal range in 40% of patients with proved primary hyperparathyroidism when the "anti N" system was used whereas this occurred in only 10% of patients when the "anti C" system was used.

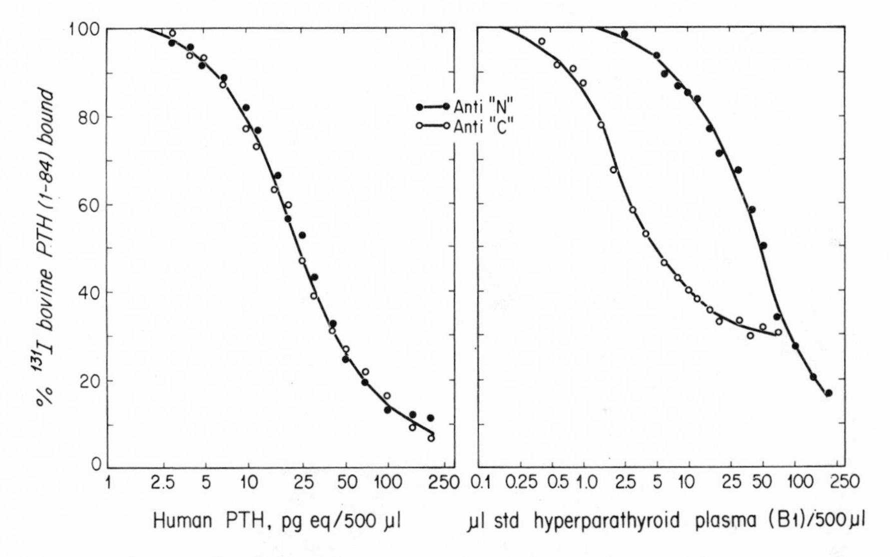

Figure 11. Standard curves demonstrating the relative reactivities of GP 1M (anti-C) antiserum (o-o) and CH 14M (anti-N) antiserum (●-●) with human adenoma PTH (1-84) (Left) and with the PTH in hyperparathyroid plasma (Right). (From Arnaud CD, Goldsmith RS, Bordier PJ, Sizemore GW: Influence of immunoheterogeneity of circulating PTH on results of radioimmunoassay of serum parathyroid hormone in man. Am. J. Med. [in press]. By permission of Dun·Donnelley Publishing Corporation.)

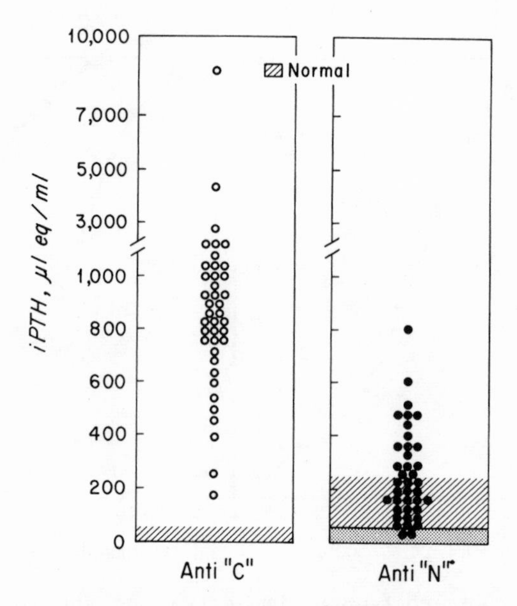

Figure 12. Serum iPTH values in patients with chronic end-stage renal failure (creatinine clearance, <5 ml/min) with GP 1M (anti-C) and CH 14M (anti-N) assays. (From Arnaud CD, Goldsmith RS, Bordier PJ, Sizemore GW: Influence of immunoheterogeneity of circulating PTH on results of radioimmunoassay of serum parathyroid hormone in man. Am. J. Med. [in press]. By permission of Dun·Donnelley Publishing Corporation.)

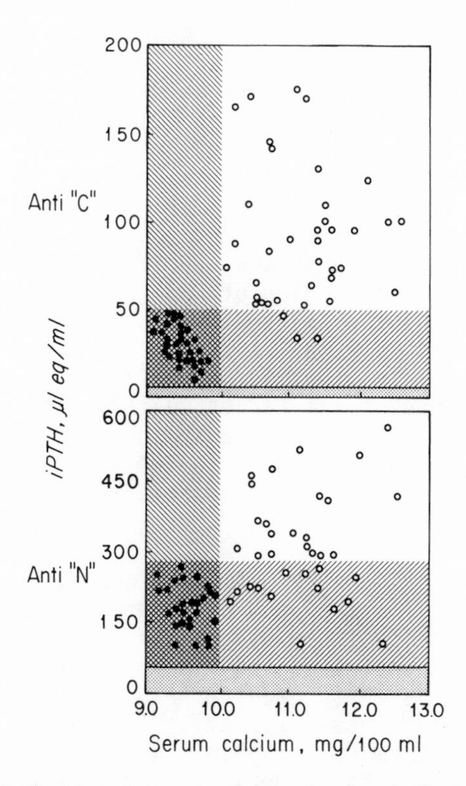

Figure 13. Relationships between serum calcium and serum iPTH values in normals (•-•) and patients with primary hyperparathyroidism (o-o), using GP 1M (anti-C) and CH 14M (anti-N) assays. (From Arnaud CD, Goldsmith RS, Bordier PJ, Sizemore GW: Influence of immunoheterogeneity of circulating PTH on results of radioimmunoassay of serum parathyroid hormone in man. Am. J. Med. [in press]. By permission of Dun·Donnelley Publishing Corporation.)

The plots in Figure 14 show that, in patients with primary hyperparathyroidism, osteoclast number increases uniformly with increases in serum iPTH as assayed with the "anti C" assay system whereas this relationship is less consistent with serum iPTH values obtained with the "anti N" assay system. Most important, however, is that there are significant increases in osteoclast number in five patients who had normal serum iPTH values as measured with the "anti N" assay system. In sharp contrast, serum iPTH assayed with the "anti C" system was consistently increased in all patients, even in those who had normal numbers of osteoclasts.

However incongruous these data appear, they clearly show that serum assays that are specific for the biologically inactive COOH-terminal region of PTH (compared with assays specific for the biologically active NH_2-terminal region) reflect better not only the state of chronic parathyroid hyperfunction but also the biologic effects of excess circulating PTH on one of its target cells (osteoclasts). The explanations for these phenomena are not clear but are probably related, at least in part, to the apparent long survival times of COOH-terminal fragments relative to PTH 1-84 in serum.[12,16,41]

In the case of parathyroid function, measurement of serum concentrations of PTH 1-84 and COOH-terminal fragments in serum (as was done by the "anti C" assay system) probably is a better index of integrated PTH secretion over the previous hours than is measurement of PTH 1-84 alone (as was done by the "anti N" assay system). This would be true whether the COOH-terminal fragments were generated by the degradation of PTH 1-84 in peripheral organs or by parathyroid tissue itself.

In the case of biologic effects, COOH-terminal fragments of PTH 1-84 might be generated and released into the circulation after binding of the whole molecule to target cell receptors, and the measurement of these fragments in serum could be an indirect index of the size and activity of the pool of PTH receptors. Preliminary evidence in support of this latter mechanism has been obtained by Di Bella and associates[42] in their studies of receptors of PTH in isolated bovine renal cortical membranes.

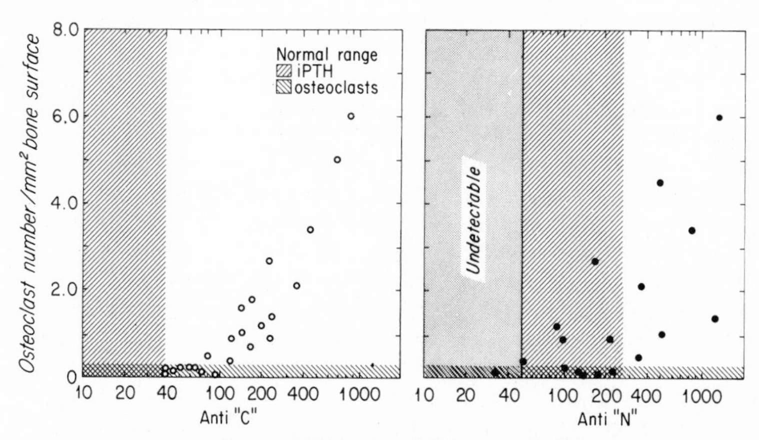

Figure 14. Relationships between serum iPTH values and osteoclast counts in bone biopsy specimens obtained from patients with primary hyperparathyroidism, using GP 1M (anti-C) (o-o) and CH 14M (anti-N) (•-•) assays. For GP 1M, $\underline{r}$ = 0.963 ($\underline{P}$ = <0.0001); for CH 14M, $\underline{r}$ = 0.669 ($\underline{P}$ <0.01). (From Arnaud CD, Goldsmith RS, Bordier PJ, Sizemore GW: Influence of immunoheterogeneity of circulating PTH on results of radioimmunoassay of serum parathyroid hormone in man. Am. J. Med. [in press]. By permission of Dun·Donnelley Publishing Corporation.)

The emphasis we have placed on the superiority of COOH-terminal specific assays of PTH in assessing chronic parathyroid hyperfunction should not be construed as a condemnation of NH_2-terminal specific assays. The COOH-terminal specific assays are much less useful than the NH_2-terminal specific assays in assessing acute changes in parathyroid function (as during induced hypercalcemia or hypocalcemia) and in demonstrating differences in iPTH in sera obtained during attempts to localize parathyroid lesions by differential catheterization of veins in the neck and mediastinum. This is because the major species of PTH secreted by parathyroid tissue is PTH 1-84 and it has a short survival in the circulation. The NH_2-terminal specific assays measure PTH 1-84 almost exclusively and essentially exclude the high "background" levels of COOH-terminal fragments (long survival time) that invariably are present in peripheral serum and may "contaminate" serum samples obtained from neck or mediastinal venous drainages. It therefore is logical to expect that NH_2-terminal specific assays would be particularly useful in the performance of these tests. Although not presented in this paper, systematic studies comparing the "anti C" and "anti N" assays systems in our laboratory strongly support these notions[40] (unpublished data).

COMMENT

The elucidation of the amino acid sequence of the biologically active region of the human PTH will now permit detailed chemical and synthetic studies on the specific amino acid residues that are required for the biologic activity of the human hormone. Also, studies will now be required to evaluate the biologic and immunologic significance of the change from a glutamic acid to a glutamine residue at position 22 in the biologically active NH_2-terminal region of the bovine and procine hormones.

We believe that the data we have presented concerning the practical consequences of the immunoheterogeneity of PTH in the circulation of hyperparathyroid man resolve, to some extent, the disagreements that have arisen about the separation of normal subjects from patients with hyperparathyroidism and about the regulation of PTH secretion in this disease. It is clear that the key variable responsible for the discordant results is antiserum specificity. Measurements of serum iPTH in both chronic and acute studies will differ, depending on this variable, not only because of the presence of multiple molecular species of PTH in the peripheral circulation but also because the ratios of the concentrations of these forms are different, depending on their metabolic turnover rates. These problems need not detract from the utility of the radioimmunoassay of PTH as an investigative and diagnostic tool. However, they dictate that knowledge of the specificity of an individual assay is essential.

ACKNOWLEDGMENT

The work on the amino acid sequence of human PTH could not have been done without the generous donations of human parathyroid tissue to us by more than 150 physicians from 12 different nations. We thank our colleagues Ms. Rosemary Ronan and Drs. Thomas Fairwell, Glen W. Sizemore, and Francis P. Di Bella who actively contributed to the work on the amino acid sequence of human PTH and Dr. Werner Rittel whose group at the Ciba-Geigy Co. synthesized human PTH 1-34. Dr. E. Travis Littledike provided the porcine parathyroid hormone for sequence analysis. Dr. Ralph S. Goldsmith was actively involved in the work on the application of region-specific radioimmunoassays of PTH to the study of parathyroid function in hyperparathyroid patients. We are especially grateful to Dr. Philippe J. Bordier who provided the osteoclast counts and serum for PTH analysis from his patients in Paris. Dr. Thomas Dousa generously carried out the renal cortical membrane adenylate cyclase assays of synthetic bovine and human PTH 1-34. The constantly superb technical assistance of Ms. Julianna Gilkinson, Ms. Judith A. Larsen, Ms. Kathleen A. Safford, and Ms. Judith M. Verheyden is greatly appreciated.

REFERENCES

1. Hamilton, J. W., MacGregor, R. R., Chu, L. L. H., Cohn, D. V.: The isolation and partial purification of a non-parathyroid hormone calcemic fraction from bovine parathyroid glands. Endocrinology, 89:1440-1447, 1971.

2. Cohn, D. V., MacGregor, R. R., Chu, L. L. H., Hamilton, J. W.: Studies on the biosynthesis in vitro of parathyroid hormone and other calcemic polypeptides of the parathyroid gland. Excerpta Medica International Congress Series No. 243, 1971, pp. 173-182.

3. Cohn, D. V., MacGregor, R. R., Chu, L. L. H., Kimmel, J. R., Hamilton, J. W.: Calcemic fraction-A: biosynthetic peptide precursor of parathyroid hormone. Proc. National Acad. Sci. U. S. A., 69: 1521-1525, 1972.

4. Kemper, B., Habener, J. F., Potts, J. T., Jr., Rich, A.: Proparathyroid hormone: identification of a biosynthetic precursor to parathyroid hormone. Proc. National Acad. Sci. U. S. A., 69:643-647, 1972.

5. Hamilton, J. W., Niall, H. D., Keutmann, H. T., Potts, J. T., Jr., Cohn, D. V.: Amino terminal sequence of bovine proparathyroid hormone (calcemic fraction A) (abstract). Fed. Proc. 32:269, 1973.

6. Habener, J. F., Kemper, B., Potts, J. T., Jr., Rich, A.: Bovine proparathyroid hormone: structural analysis of radioactive peptides formed by limited cleavage. Endocrinology, 92:219-226, 1973.

7. Cohn, D. V., MacGregor, R. R., Chu, L. L. H., Huang, D. W. Y., Anast, C. S., Hamilton, J. W.: Biosynthesis of proparathyroid hormone: chemistry, physiology, and role of calcium in regulation. Am. J. Med. (in press).

8. Habener, J. F., Powell, D., Murray, T. M., Mayer, G. P., Potts, J. T., Jr.: Parathyroid hormone: secretion and metabolism in vivo. Proc. National Acad. Sci. U. S. A., 68:2986-2991, 1971.

9. Berson, S. A., Yalow, R. S.: Immunochemical heterogeneity of parathyroid hormone in plasma. J. Clin. Endocrinol. Metab., 28:1037-1047, 1968.

10. Arnaud, C. D., Tsao, H. S., Oldham, S. B.: Native human parathyroid hormone: an immunochemical investigation. Proc. National Acad. Sci., U. S. A., 67:415-422, 1970.

11. Arnaud, C. D., Sizemore, G. W., Oldham, S. B., Fischer, J. A., Tsao, H. S., Littledike, E. T.: Human parathyroid hormone: glandular and secreted molecular species. Am. J. Med. 50:630-638, 1971.

12. Canterbury, J. M., Reiss, E.: Multiple immunoreactive molecular forms of parathyroid hormone in human serum. Proc. Soc. Exp. Biol. Med. 140:1393-1398, 1972.

13. Habener, J. F., Segre, G. V., Powell, D., Murray, T. M., Potts J. T., Jr.: Immunoreactive parathyroid hormone in circulation of man (letter to the editor). Nature [New Biol.] 238:152-154, 1972.

14. Segre, G. V., Habener, J. F., Powell, D., Tregear, G. W. Potts, J. T., Jr.: Parathyroid hormone in human plasma: immunochemical characterization and biological implications. J. Clin. Invest. 51:3163-3172, 1972.

15. Arnaud, C. D., Goldsmith, R. S., Sizemore G. W., Oldham, S. B., Bischoff, J., Larsen, J. A., Bordier, P: Studies on characterization of human parathyroid hormone in hyperparathyroid serum: practical considerations. Excerpta Medica International Congress Series No. 270, 1973, pp. 281-290.

16. Silverman, R., Yalow, R. S.: Heterogeneity of parathyroid hormone: clinical and physiologic implications. J. Clin. Invest. 52:1958-1971, 1973.

17. Arnaud, C. D.: Parathyroid hormone: coming of age in clinical medicine. Am. J. Med. 55:577-581, 1973.

18. Arnaud, C. D., Goldsmith, R. S., Bordier, P. J., Sizemore, G. W.: Influence of immunoheterogeneity of circulating PTH on results of radioimmunoassay of serum parathyroid hormone in man. Am. J. Med. (in press).

19. Segre, G. V., Niall, H. D., Habener, J. F., Potts, J. T., Jr.: Metabolism of parathyroid hormone: physiologic and clinical significance. Am. J. Med. (in press).

20. Canterbury, J. M., Levey, G. S., Reiss, E.: Activation of renal cortical adenylate cyclase by circulating immunoreactive parathyroid hormone fragments. J. Clin. Invest. 52:524-527, 1973.

21. Rasmussen H., Sze, Y. L., Young, R.: Further studies on the isolation and characterization of parathyroid polypeptides. J. Biol. Chem. 239:2852-2857, 1964.

22. Brewer, H. B., Jr., Fairwell, T., Ronan R., Sizemore, G. W., Arnaud, C. D.: Human parathyroid hormone: amino-acid sequence of the amino-terminal residues 1-34. Proc. National Acad. Sci., U. S. A., 69:3585-3588, 1972.

23. Pisano, J. J., Bronzert, T. J., Brewer, H. B., Jr.: Advances in the gas chromatographic analysis of amino acid phenyl- and methylthiohydantoins. Anal. Biochem. 45:43-59, 1972.

24. Fales, H. M., Nagai, Y., Milne, G. W. A., Brewer, H. B., Jr., Bronzert, T. J., Pisano, J. J.: Use of chemical ionization mass spectrometry in analysis of amino acid phenylthiohydantoin derivatives formed during Edman degradation of proteins. Anal. Biochem. 43:288-299, 1971.

25. Fairwell, T., Brewer, H. B., Jr.: Identification by chemical ionization mass spectrometry (CI-MS) of the phenyl (PTH) and methyl (MTH) thiohydantoin and thiazolinone amino acids obtained during the automated Edman degradation of polypeptides and proteins (abstract). Fed. Proc. 32:648, 1973.

26. Brewer, H. B., Jr., Ronan, R.: Bovine parathyroid hormone: amino acid sequence. Proc. National Acad. Sci., U. S. A., 67:1862-1869, 1970.

27. Niall, H. D., Keutmann, H., Sauer, R., Hogan, M., Dawson, B., Aurbach, G., Potts, J., Jr.: The amino acid sequence of bovine parathyroid hormone. I. Hoppe Seylers Z. Physiol. Chem., 351:1586-1588, 1970.

28. Brewer, H. B., Jr., Fairwell, T., Rittel, W., Littledike, T., Arnaud, C. D.: Recent studies on the chemistry of human, bovine, and porcine parathyroid hormones. Am. J. Med. (in press).

29. O'Riordan, J. H., Woodhead, J. S., Robinson, C. J., Parsons, J. A., Keutmann, H., Niall, H., Potts, J. T.: Structure-function studies in parathyroid hormone. Proc. R. Soc. Med., 64:1263-1265, 1971.

30. Tregear, G. W., van Rietschoten, J., Greene, E., Keutmann, H. T., Niall, H. D., Reit, B, Parsons, J. A., Potts, J. T., Jr.: Bovine parathyroid hormone: minimum chain length of synthetic peptide required for biological activity. Endocrinology 93:1349-1353, 1973.

31. Potts, J. T., Jr., Niall, H. D., Tregear, G. W., van Rietschoten, J., Habener, J. F., Segre, G. V., Keutmann, H. T.: Chemical and biologic studies of proparathyroid hormone and parathyroid hormone: analysis of hormone biosynthesis and metabolism. Mt. Sinai, J. Med., N. Y., 40:448-461, 1973.

32. Andreatta, R. H., Hartmann, A., Jöhl, A., Kamber, B., Maier, R., Riniker, B., Rittel, W., Sieber, P.: Synthese der Sequenz 1-34 von menschilichem Parat-hormon. Helv Chim Acta, 56:470-473, 1973.

33. Reiss, E., Canterbury, J. M.: A radioimmunoassay for parathyroid hormone in man. Proc. Soc. Exp. Biol. Med., 128:501-504, 1968.

34. Berson, S. A., Yalow, R. S.: Parathyroid hormone in plasma in adenomatous hyperparathyroidism, uremia, and bronchogenic carcinoma. Science, 154:907-909, 1966.

35. Reiss, E., Canterbury, J. M.: Primary hyperparathyroidism: application of radioimmunoassay to differentiation of adenoma and hyperplasia and to preoperative localization of hyperfunctioning parathyroid glands. N. Engl. J. Med. 280:1381-1385, 1969.

36. Potts, J. T., Murray, T. M., Peacock, M., Niall, H. D., Tregear, G. W., Keutmann, H. T., Powell, D., Deftos, L. J.: Parathyroid hormone: sequence, synthesis, immunoassay studies. Am. J. Med., 50:639-649, 1971.

37. Potts, J. T., Jr., Keutmann, H. T., Niall, H. D., Tregear, G. W., Habener, J. F., O'Riordan, J. L. H., Murray, T. M., Powell, D., Aurbach, G. D.: Parathyroid hormone: chemical and immunochemical studies of the active molecular species. In Endocrinology, 1971, (Proceedings of the Third International Symposium). Edited by S. Taylor. London, William Heinemann, 1972, pp. 333-349.

38. Arnaud, C. D.: Immunochemical heterogeneity of circulating parathyroid hormone in man: sequel to an original observation by Berson and Yalow. Mt. Siani, J. Med., N. Y., 40:422-432, 1973.

39. Arnaud, C. D., Tsao, H. S., Littledike, T.: Radioimmunoassay of human parathyroid hormone in serum. J. Clin. Invest., 50:21-34, 1971.

40. Benson, R. C., Jr., Riggs, B. L., Pickard, B. M., Arnaud, C. D.: Immunoreactive forms of circulating parathyroid hormone in primary and ectopic hyperparathyroidism. J. Clin. Invest. (in press).

41. Goldsmith, R. S., Furszyfer, J., Johnson, W. J., Fournier, A.E., Sizemore, G.W., Arnaud, C.D.: Etiology of hyperparathyroidism and bone disease during chronic hemodialysis. III. Evaluation of parathyroid suppressibility. J. Clin. Invest., 52:173-180, 1973.

42. DiBella, F. P., Dousa, T. P., Miller, S. S., Arnaud, C. D.: Parathyroid hormone receptors of renal cortex: specific binding of bioligically active, ^{125}I-labeled hormone and relationship to adenylate cyclase activation. Proc. National Acad. Sci., U. S. A., 71:723-726, 1974.

This investigation was supported in part by Research Grants AM-12302 from the National Institutes of Health, Public Health Service, and by a grant from the Mayo Foundation.

Reprint requests and correspondence should be addressed to Dr. Arnaud, Mayo Clinic, Rochester, Minnesota 55901.

Read at the X European Symposium on Calcified Tissue, Hamburg, Germany, September 17 to 21, 1973.

IN VITRO RADIOISOTOPIC METHODS FOR CLINICAL EVALUATION OF VITAMIN B_{12} AND FOLIC ACID METABOLISM

Sheldon P. Rothenberg, M.D.

New York Medical College
1249 Fifth Avenue
New York, N.Y. 10029

Many important physiologic processes are mediated by substances present in concentrations too low to be measured with instrumentation dependent on colorimetric or fluorometric reaction systems. Consequently, the assay procedures for many hormones, enzymes, metabolic substrates and vitamins was dependent for many years on biological activity where sensitivity, if not precision, was greater. However, with the availability of radioactive isotopes which could be incorporated into these substances, a significant advance had been made because they could now be monitored at lower and more physiologic concentrations. Coupled with the availability of radioactive labels was the development of the competitive radioimmunoassay for insulin by Yalow and Berson (1) which probably was the singular most important advance in the field of biomedical measurement. This technique, first exploited for the measurement of insulin, has now been extended for the measurement of other hormones, both peptide and non-peptide, other types of proteins, vitamins and drugs, using, in addition to specific antibodies, other macromolecular binding ligands or receptor sites obtained from tissues or biological fluids.

One of the earliest applications of the principle of competitive ligand binding for the radioassay of substances other than peptide hormones was for the measurement of vitamin B_{12}. This occurred for two reasons. First, vitamin B_{12} labeled with isotopic cobalt, a natural component of the vitamin, became available; and second, a natural binder of B_{12}, gastric intrinsic factor, could be used as a ready source of binding ligand (2). For chronologic development of the B_{12} radioassay and summary of methodology and binding ligands, the reader is referred to a recent review (3).

Several natural binding ligands for vitamin B_{12} are available and their high binding affinity is particularly suited for the sensitivity required to measure picogram quantities of this vitamin. Intrinsic factor (IF), a protein secreted by the stomach, is a natural binder of B_{12} which facilitates the intestinal absorption of the vitamin. This protein has been partially purified from animal sources, particularly the hog, and it is now probably the most commonly used binding ligand for the radioassay of B_{12}. Serum and saliva also contain binding proteins of this vitamin and these have also been used as the source of reacting ligand (4, 5, 6, 7).

The fundamental reaction for the B_{12} radioassay procedure is the same as that for all competitive ligand binding systems and is shown in the equation below where B_{12}^* and B_{12} are isotopic and unlabeled vitamin, respectively, and Bl is the binding ligand (either IF or serum B_{12} binders).

$$B_{12}^* + B_{12} + Bl \rightleftharpoons B_{12}^* - Bl + B_{12} - Bl$$

When this reaction is stopped, which may or may not be at equilibrium, B_{12}^* bound to binding ligand (Bl) will vary inversely to the concentration of non-isotopic B_{12}. It is necessary to separate the free and ligand bound forms of B_{12} (isotopic and non-isotopic) by some physico-chemical method in order to determine the fraction of each. For this purpose several procedures have been described such as $ZnSO_4$ - $Ba(OH)_2$ precipitation (2) coated charcoal (9) and ion exchanges (6). More recently, IF insolubilized on carrier molecules has been used and this permits easy separation of free and bound B_{12} and also eliminates one step in the procedure (see Ref. 3).

After determining the bound and free fractions of B_{12}*, a standard curve can be constructed correlating the fraction of bound B_{12}* to the concentration of unlabeled B_{12} in the reaction mixture. When solutions containing an unknown concentration of B_{12} are similarly assayed under the same experimental conditions, using the same concentration of isotopic B_{12} and binding ligand, the concentration of the vitamin in the solution can be obtained by referring the experimentally determined value of B_{12}* bound to this standard curve.

Although intrinsic factor is the binding ligand most commonly used in the B_{12} radioassay, for reproducible dose-response sensitivity beginning at 10 to 20 pg we have found that the B_{12} binding protein(s) which is contained in the serum of patients with chronic myelogenous leukemia is far more reliable (8). The B_{12} binder(s) in such serum has an extremely high affinity for B_{12} with virtually no dissociation under the conditions of the radioassay and, in addition, it is more stable than intrinsic factor. In the dose-response range between 0 and 200 pg, the reciprocal of the fraction bound plotted against the concentration of non-isotopic standard B_{12} is linear and it is easily adapted to desk-top computerization. We now routinely use 2.5% charcoal in 0.25% dextran solution to separate bound and free fractions of the vitamin. An example of a standard curve which we routinely obtain is shown in Figure 1. The stability of the binder either in the whole serum or following fractionation with 2 M $(NH_4)_2SO_4$ to remove other B_{12} binders, permits reliable duplication of the standard curve so that only one curve need be established with each lot of binder preparation.* In this way single samples of serum can be assayed without the need to construct a new standard curve. The reproducibility of the B_{12} value assayed repetitively against the same standard curve is $\pm$ 10 to 15%.

*A simple technique to improve the homogeneity of B_{12} binders in such serum is to treat it with 150 mg of dry charcoal per ml to remove the low molecular weight beta binder called TC-II.

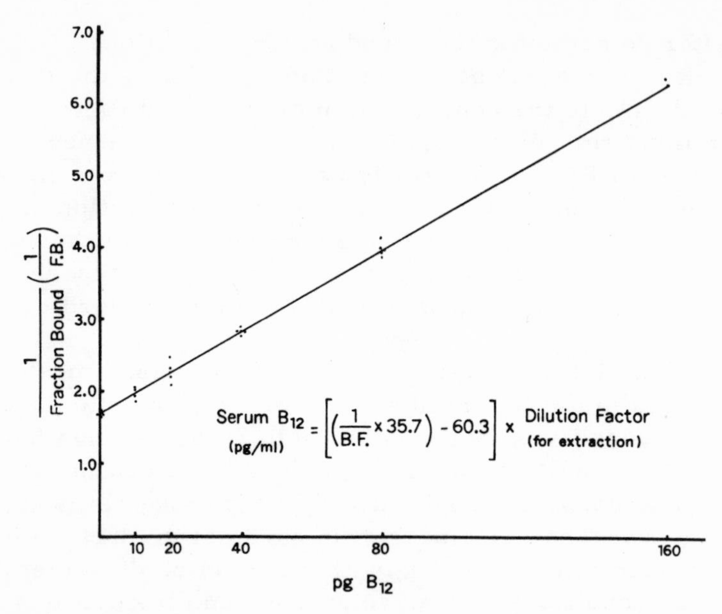

Figure 1. Standard curve obtained using TC-I containing serum. The reciprocal of the fraction bound (1/F.B.) is plotted as a function of the unlabeled B_{12} in the reaction mixtures. Each point at 10, 20, 40, 80, and 160 pg was determined four different times to show the scatter around the standard line.

Vitamin B_{12} in blood and in most tissues is bound to several different proteins. It is necessary, therefore, to dissociate the vitamin in order to assay it by the competitive ligand binding system. Most laboratories boil the serum (or tissue extract) at a pH which will precipitate most proteins. We have obtained the best results with dissociation of the greatest amount of endogenous B_{12} and removing or denaturing most of the serum binders by first boiling one volume of serum in 2 volumes of 0.68 M acetate Ringer's buffer, pH 4.3, containing 10 μg per ml sodium cyanide for 15 minutes followed by 15 minutes of additional boiling after adding 1 ml of 0.3 N NaOH to raise the pH to 5.6 (8). The cyanide is an important constituent of the extraction buffer and without it lower yields of B_{12} are obtained. Other methods for extracting bound B_{12} usually with boiling at a single pH, or acid inactivation of serum binders with dissociation of endogenous B_{12} without precipitation of serum proteins has also been described (see review, ref. 2).

The normal range of serum B_{12} values using this method is 200 to 800 pg per ml. Patients with leukemia, particularly chronic myelogenous leukemia, contain very high values, while patients with impaired B_{12} absorption such as pernicious anemia contain less than 200 pg/ml and frequently less than 100 pg/ml.

The clinical value of serum B_{12} determination is summarized in Table I. With very few exceptions, it can be stated that when the serum B_{12} is low, the patient has vitamin B_{12} deficiency. The causes for such deficiency must then be determined by appropriate studies. Specific diagnostic procedures are available to help the physician determine whether such deficiency is due to nutritional inadequacy (which is very rare), impaired intrinsic factor secretion by the stomach, or impaired intestinal absorption.

TABLE I
Clinical Value of Serum B_{12} Concentration

I. Low B_{12} Concentration: Less than 200 pg/ml
   1. Nutritional deficiency
   2. Pernicious anemia
   3. Impaired intestinal absorption
   4. Pregnancy

II. High B_{12} Concentration
   1. Liver disease
   2. Myeloproliferative syndrome
   3. B_{12} administration

High serum B_{12} concentrations can also be diagnostically helpful. If not due to recent injection of B_{12}, the most common cause of this finding is liver disease or a group of diseases called the myeloproliferative diseases, particularly chronic myelogenous leukemia. Leukemoid reactions may also elevate the serum B_{12} concentration.

The availability of isotopic B12 has also permitted studies of the serum B_{12} proteins. There is now an extensive literature on this subject and the details are beyond the scope of this discussion. The method generally applied to define these B_{12} binders is to saturate the serum (or other source of fluid) with radioactive B_{12}, remove the excess or unbound B_{12} by dialysis or coated charcoal, and then subject the bound B_{12} to procedures which will separate the proteins on the bases of molecular weight using sephadex gel filtration, or on the basis of charge using ion exchange chromatography. With gel filtration and ion exchange procedures there appears to be three proteins in serum which bind B12 and the properties are summarized in Table II. Transcobalamin has been proposed as the nomenclature for the serum B_{12} binders by Hall and Finkler (10, 11), and their studies defined many of the properties of these proteins. One binder, called transcobalamin I, is an acidic protein which binds strongly to DEAE and has a molecular weight greater than 100, 000 by gel filtration. It moves with the mobility of an α_1 globulin on electrophoresis at pH 8.6. A second binder, or transcobalamin II, elutes with dilute buffer from DEAE cellulose, has a beta mobility on electrophoresis and has a molecular weight of approximately 38, 000 by gel filtration studies. A third binder is similar in size to transcobalamin I, but elutes from DEAE cellulose and moves electrophoretically as TC-II (12, 13). This has been called transcobalamin III (14).

The nature and origin of these B_{12} binders is under intense investigation in many laboratories. The biologic function of TC-I and TC-III are unknown although endogenous serum B_{12} is carried for the most part on TC-I. TC-II remains for the most part unsaturated and binds newly absorbed B_{12} or B12 added to serum in vitro and it releases B12 for uptake by tissues. TC-III also remains unsaturated in vivo, but binds B_{12} added to serum in vitro.

Most recently Allen and Majerus have isolated from serum and granulocytes only two B_{12} binders using affinity

TABLE II
Properties of B_{12} Binding Proteins in Serum*

	TC-I	TC-II	TC-III
Mol. Wt (gel filtration)	120,000	36,000	120,000
$(NH_4)_2SO_4$ precipitation	full saturation	1/2 saturation	full saturation
Electrophoresis (pH 8.6)	α_1 globulin	β globulin	β globulin
DEAE cellulose	high affinity	low affinity	low affinity
Charcoal adsorption	no	yes	no
Source	granulocyte	liver	granulocyte

*The serum B_{12} binding proteins have recently been named transcobalamins and abbreviated TC.

chromatography (15, 16). These B_{12} binders are similar to TC-I and TC-II, but with molecular weights of 56,000 and 53,000, respectively, by sedimentation equilibrium ultracentrifugation. The higher molecular weight of TC-I by gel filtration has been attributed to its 33% carbohydrate content. Further studies are necessary to explain TC-III which has been identified by gel filtration and ion exchange chromatography, but not by affinity chromatography.

The determination of the unsaturated serum B_{12} binding capacity can be particularly helpful in certain disease states. Because the origin of TC-I and TC-III appears to be the granulocyte (17), the serum unsaturated binding capacity of these binders will have high values in disorders characterized by excessive granulopoiesis. This is particularly so in chronic granulocytic leukemia. In liver disease, on the other hand, serum B_{12} and unsaturated B_{12} binding capacity is elevated

TABLE III
Principles of Methodology to Determine Binding Capacity of Specific B_{12} Binding Proteins in Serum

I.	Total Binding Capacity: a. Saturate with $^{57}CoB_{12}$ b. Remove excess B_{12} with coated charcoal
II.	TC-II Binding Capacity: a. Treat serum with dry charcoal (150 mg/ml) b. Determine binding capacity as in I TC-II = I - II
III.	TC-I and TC-III Binding Capacity: a. Treat serum with charcoal b. Saturate with $^{57}CoB_{12}$ - remove excess with charcoal c. Pass through small DEAE cellulose column with 0.06 M phosphate buffer, pH 6.3 TC-I = radioactivity remaining on column TC-III = radioactivity eluting from column

with TC-II as the predominant binder because the liver is the likely source of this protein.

The methodology for determining the unsaturated B_{12} binding capacity for each of the three B_{12} binding proteins in serum has been based on the properties of these binders and are summarized in Table III. First, sufficient isotopic B_{12} is added to serum to saturate the binders and the unbound B_{12} is removed with coated charcoal (9), or dialysis. The isotopic B_{12} remaining is a measure of the total unsaturated B_{12} binding capacity. This procedure is then repeated on a sample of the serum which was first treated for 30 minutes with dry charcoal (150 mg per ml) to remove TC-II (13). The decrease in B_{12} bound is a measure of the TC-II unsaturated binding capacity and the amount of isotopic B_{12} remaining represents the TC-I + TC-III unsaturated binding capacity. Finally, the remaining mixture TC-I and TC-III B_{12} mixture is filtered through a small DEAE cellulose column with 0.06 M phosphate buffer, pH 6.3. This elutes the TC-III binder leaving the TC-I on the column. The radioactivity of each of the fractions is converted to B_{12} concentration using the specific activity of the tracer B_{12}.

When serum is obtained to determine the unsaturated B_{12} binding capacity, serum binder(s) may be released from the granulocyte during clotting (18). If heparinized plasma is used, lithium contained in some commercial heparin may stimulate the release of binder from granulocytes (19). It is recommended, therefore, that blood for B_{12} binding capacity be collected in fluoride which prevents the metabolic activity resulting in in vitro release of B_{12} binders (18).

A very helpful test in patients with low serum B_{12} is the detection of serum antibodies to intrinsic factor for, if present, it indicates that the patient has pernicious anemia. Depending on the assay system used, such antibodies can be found in 50 to 75% of patients with pernicious anemia. Actually, two types of antibody to IF have been identified in these patients (20, 21). These are shown diagrammatically in Figure 2. When isotopic B_{12} (B_{12}*) is added to a solution containing IF, it is quickly bound. One type of antibody called type I or blocking antibody, prevents the binding of B_{12}* by IF if added to the reaction before the antibody containing serum. The amount of radioactivity remaining unbound, therefore, by any method of determination such as dialysis, coated charcoal, or zinc-sulfate $Ba(OH)_2$ precipitation is a measure of the blocking effect of the antiserum.

The second type of antibody attaches to the part of the IF molecule which is distant from the B_{12} binding site. Therefore, it does not block the coupling of B_{12} yet it attaches itself to IF molecule and has been called type II or binding antibody. This antibody is detected by determining changes

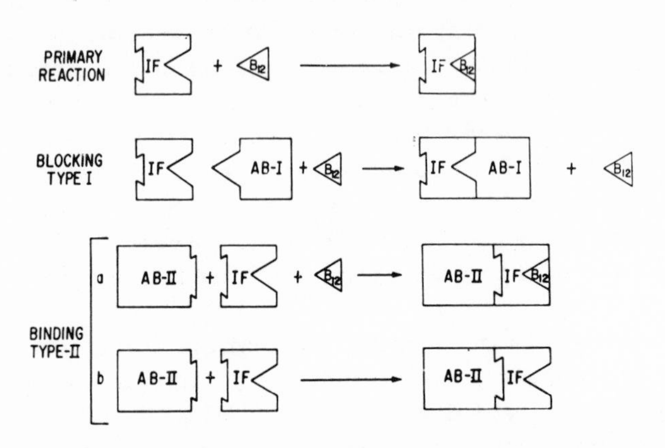

Figure 2. Antibody to intrinsic factor. One type of antibody blocks the binding of B_{12} to intrinsic factor (type I) and the second type of antibody binds the intrinsic factor - B_{12} complex (type II).

in physico-chemical properties of IF-B_{12} as a result of binding to the immunoglobulins.

The methods used for the detection of each type of antibody have generally been different and it has, therefore, been necessary to test for each separately. Because of the relatively high incidence of B_{12} deficiency in our patient population, and the frequent requirement for testing for these antibodies, we developed a radioimmunoassay which can be used to detect simultaneously the presence of blocking and binding antibody in serum (22). The details are summarized in Table IV. To test for blocking antibody, normal human gastric juice diluted so that the IF binds 60 to 70% of 120 pg of $^{57}CoB_{12}$ is incubated with the test serum for 5 minutes prior to the addition of the tracer B_{12}. Upon the addition of the $^{57}CoB_{12}$, little or no radioactivity will bind to the IF of the gastric juice if the serum contains blocking antibody. To determine how much of the tracer B_{12} was bound to the IF, antiserum (previously saturated with 2 ng per ml of crystalline B_{12}) with binding antibody to IF is ad-

TABLE IV
Procedures for the Immunoassay of Serum-Blocking and -Binding Anti-IF Antibodies*

Blocking		Binding	
Reaction mixture volume (ml)			
Tris buffer	1.2	Tris buffer	1.2
Diluted NHGJ	0.1	Diluted NHGJ	0.1
Test sample+	0.1	Tracer B_{12} (120 pg)	0.1
5 Minute Incubation			
Tracer B_{12} (120 pg)	0.1	Test sample+	0.1
5 Minute Incubation			
Binding anti-IF serum	0.5	Normal serum	0.5
5 Minute Incubation			
30% Na_2SO_4	2.0	30% Na_2SO_4	2.0
15 to 20 Minute Incubation			
Count radioactivity in supernate		Count radioactivity in supernate	

*All incubations were at room temperature.
+Serum or reconstituted precipitate from gastric juice.

ded to the reaction mixture. At a concentration of 15% Na_2SO_4, which is achieved by adding an equal volume of 30% Na_2SO_4, the antibody-IF complex precipitates while free IF remains soluble. The tracer B_{12} will co-precipitate with the antibody-IF complex if the test serum did not contain blocking antibody. If the $^{57}CoB_{12}$ did not precipitate, however, then the test serum did contain blocking antibody.

Only slight modification is needed to determine the presence of binding antibody. In this instance, the diluted gastric juice is incubated first with the $^{57}CoB_{12}$ tracer to permit the IF to be "isotopically labeled", and then with the

TABLE V
Procedure for the Immunoassay of Intrinsic Factor in Gastric Juice*

Reaction Mixture	Volume (ml)
0.02 M Tris buffer, pH 7.2	1.3
Gastric juice	0.1
Tracer B_{12} (7.5 ng)	0.1
5 Minute Incubation	
Binding anti-IF serum	0.5
5 Minute Incubation	
30% Na_2SO_4	2.0
15 to 20 Minute Incubation	
Count radioactivity in supernate+	

*All incubations were at room temperature.
+Calculations of IF binding capacity:
ng B_{12}/ml gastric juice = % $^{57}CoB_{12}$ ppted x 7.5 x 10

test serum. If binding antibody is present, the whole complex of antibody-IF-$^{57}CoB_{12}$ will then precipitate at 15% Na_2SO_4. In this instance, B_{12} saturated normal serum is added prior to the Na_2SO_4 only to supply sufficient carrier immunoglobulin protein.

The binding of IF by anti-IF antibody has also been used in our laboratory to measure the content of IF in gastric juice (22). This reaction system for this radioimmunoassay is summarized in Table V. An aliquot of gastric juice is first incubated with a saturating concentration of $^{57}CoB_{12}$, followed in a few minutes by antiserum containing an excess of binding type antibody. The immunoglobulins are then precipitated at 15% Na_2SO_4. Only $^{57}CoB_{12}$ bound to IF will coprecipitate and the concentration of IF is expressed in terms of ng of B_{12} bound per ml gastric juice. Normal gastric juice collected 15 minutes after histamine stimulation con-

TABLE VI
Clinical Value of Antibodies to Intrinsic Factor

1. Specifically diagnostic of pernicious anemia when associated with a low serum B_{12} concentration.

2. Positive test avoids necessity for gastric analysis.

3. Positive test avoids necessity for radio-B_{12} absorption study to diagnose pernicious anemia.

4. Helpful in screening subjects with:
   a. hyperthyroidism
   b. hypothyroidism
   c. adrenal insufficiency
   d. mental disorders
   e. relatives of patients with pernicious anemia

5. Assay of IF in gastric juice.
   a. B_{12} deficiency negative for serum anti-IF antibodies
   b. gastritis
   c. post gastrectomy
   d. carcinoma of stomach

tains sufficient IF to bind more than 25 ng of B_{12} per ml. Gastric juice from patients with PA usually binds less than 5 ng of B_{12} per ml and frequently zero B_{12}. Intermediate values for B_{12} binding indicate some degree of gastric pathology.

There are a number of instances where the demonstration of serum antibodies to intrinsic factor and the determination of IF in gastric juice can be clinically helpful. These are summarized in Table VI.

Whereas the principle of competitive inhibition was a logical application for the radioassay of vitamin B_{12} because of the ready availability of natural macromolecular binders of this vitamin, similar application for the radioassay of folic acid (PGA) was slower to develop for two reasons. First, and most important, was that little was known about natural folate binders. Second, no isotopic form was available until tritiated PGA was synthesized and the early production runs were of poor purity (23) and low specific activity. Recent preparations of ^{3}H-PGA have improved in quality. In addition, a number of binders of PGA and folate analogues have been identified in milk (24), kidney extracts (25), leukemic cells (26), body fluids (27), and bacteria (28). It has also been possible to couple PGA to immunogenic macromolecules and obtain binding antibodies following appropriate immunization of rabbits (29) and guinea pigs (30).

The first reacting ligand we used for the radioassay of PGA by competitive inhibition was the enzyme folate reductase (also called dihydrofolate reductase) (31). This enzyme reduces PGA to tetrahydrofolate (FH_4) with NADPH as cofactor. If the enzyme concentration is such that the reaction proceeds at maximal velocity at the concentration of the tracer ^{3}H-PGA (approximately 0.57 p moles), then the ^{3}H-PGA reduced to tetrahydrofolate will be inversely related to the unlabeled PGA added to the reaction mixture. The PGA is separated from generated tetrahydrofolate by precipitation with excess PGA and $ZnSO_4$ at the completion of the reaction (32). A standard curve can be graphed by plotting the reciprocal of the fraction of ^{3}H-PGA reduced to FH_4 as a function of the concentration of unlabeled PGA.

Unfortunately, the instability of the enzyme at low concentrations limited its usefulness as a radioassay for physiologic concentrations of PGA, but we have found it quite applicable as an extremely sensitive radioenzymatic assay for methotrexate, a folate analogue which is a stoichiometric inhibitor of the enzyme with very high affinity (31).

Another step in the development of a radioassay for folate was taken when Ricker and Stollar reported that antibodies to PGA could be obtained by immunizing rabbits with the vitamin coupled to methylated bovine albumin (29). We confirmed these findings (30) and were then able to use the antibody for a radioimmunoassay for PGA (33). The antibody has determinants only for PGA and not the reduced folates such as N^5-methyltetrahydrofolate and N^5-formyltetrahydrofolate. Since the major natural folate in serum is methyltetrahydrofolate, which is present normally in concentrations greater than 4 ng per ml, we expected and, indeed, found that the concentration of unreduced immunoreactive folate (IFA) in serum was quite low but nevertheless measurable. In order to measure this folate, the serum had to be extracted without ascorbate, otherwise the folate would be reduced to a non-reactive form. Patients who were folate deficient had extremely low levels of this IFA in serum. Following oral administration of PGA, serum IFA increased considerably.

The most recent and practical ligand binding radioassay for folate has been developed using a folate binder in milk. Waxman and coworkers (34) used whole milk as the source of binder, but we have prepared a partially purified binder from milk from which a major fraction of endogenous folate has been removed (35). This binder has significantly greater affinity for PGA than for reduced folate which presents a problem using a competitive reaction system since only isotopic PGA is available as tracer. Consequently, with methyltetrahydrofolate as the unlabeled standard competing against ^{3}H-PGA for binder ligand, it is not possible to obtain a dose-response curve with sensitivity below 0.5 ng using the classical competitive reaction system where binder is added last to the reaction system. However, this problem is circumvented with a sequential non-competitive system where binder, in the first phase, is incubated with the methyltetrahydrofolate standard (or serum to be assayed) and then, in the second phase, with the ^{3}H-PGA. The tracer PGA in a sense is then really titrating the binding sites on the binding ligand unoccupied by the methyltetrahydrofolate. To improve sensitivity even more, dissociation of the binder-methyltetrahydrofolate complex is reduced by lowering the temperature of the reaction before the addition of ^{3}H-PGA. Free and bound substrates are separated using a suspension of 5% charcoal in 0.5% dextran (mol. wt. 40, 000).

With this two phase radioassay system, we obtain a dose-response curve which begins at 10 to 20 pg of methyltetrahydrofolate. This sensitivity is important because it is then only necessary to assay 25 μl of serum and this minimizes the effects of endogenous folate binders on the assay system. This is important because some serums contain a binder with determinants for PGA but not methyltetrahydrofolate. If a larger volume of serum is assayed, therefore, this endogenous binder can be formidable since it will selectively reduce this concentration of ^{3}H-PGA. Where the concentration of this endogenous binder is very high, as in some instances of folate deficiency, leukemia, or pregnancy (36), it will be necessary to extract the serum in order to assay the folate.

Serum folate concentration by this two phase system is lower than the concentration by microbiological assay using Lactobacillus casei, but normal subjects are easily separated from folate deficient patients. The higher values by L. casei suggest that some folate is present in serum in a form which is not reactive with the milk binder preparation.

Recently, Waxman and Schreiber reported that commercial beta lactoglobulin can be used as a source of the binding ligand (37). They used a competitive reacting system and because the dose-response curve began at 0.5 ng of methyltetrahydrofolate, they assayed 0.4 ml of serum. With this volume of serum, the endogenous binder of PGA can interfere to a variable extent by selectively binding the tracer.

A more recent radioisotopic assay for serum folate has been described by Kamen and Caston (25) using an extract of hog kidney as a source of binding ligand. This binder has determinants for both PGA and methyltetrahydrofolate and a dose-response curve beginning at approximately 50 pg could be obtained using a competitive reaction system (binder added to mixture of ^{3}H-PGA and methyltetrahydrofolate). This could prove to be an advantage over the folate binder in milk.

The radioassay for folate which we described (35) has now been used to measure the folate in red cells (38). For this purpose, we first lyse the red cells contained in 0.1 ml of whole blood in 25 volumes of water. Following the addition of 25 volumes of Ringer's-ascorbate solution, pH 4.9, the lysed cells are incubated for two hours at 37°C and the mixture then boiled for 5 minutes to deproteinate the solution. A small volume of the clear supernate is then assayed. Normal values are usually greater than 150 ng per ml erythrocytes, whereas patients with folate deficiency have values lower than this and frequently below 100 ng per ml. The incubation of the lysed red cells before boiling increases considerably the folate concentration suggesting that red cell folate is in a heterogenous form. It may be that it is bound to a macromolecular component which dissociates with incubation or the folate exists in part as a very high polyglutamate which is non-reactive with the binder until deconjugated at 37°C to oligoglutamates. We do know that hepta, tri, and diglutamates of folate are reactive with the milk folate binder.

The next developmental steps with respect to folate radioassays is the development of greater specificity. The milk binding ligand reacts with reduced methyltetrahydrofolate and unreduced PGA, thus, in effect, measuring total folate. We have just completed some work using a folate binder purified from leukemic cells which binds PGA but not the fully reduced folates. It is similar, therefore, to the antibody obtained to haptenic folate but is available in greater supply and with a consistently high binding affinity - a property difficult to obtain by immunization. A standard curve has been obtained with a dose-response sensitivity beginning at 10 to 20 pg. Extracts of serum contain a folate which can be assayed against this curve. Normal values vary from 0.5 to 1.5 ng per ml and in patients with folate deficiency, the concentration is significantly lower and not infrequently zero. We are presently investigating the significance of this unreduced folate with respect to folate metabolism under normal and perturbed conditions.

References

1. Yalow, R.S. and Berson, S.A.: J. Clin. Invest. 39:1157, 1960.

2. Rothenberg, S.P.: J. Clin. Invest. 42:1391, 1963.

3. Rothenberg, S.P.: Metab., 22:1075, 1973.

4. Barakat, R.M. and Ekins, R.P.: Blood 21:70, 1963.

5. Crossowicz, D., Sulitzeanu, D. and Merzbach, D.: Proc. Soc. Exp. Biol. Med., 109:604, 1962.

6. Frenkel, E.P., Keller, S. and McCall, M.S.: J. Lab. Clin. Med., 68:510, 1966.

7. Carmel, R. and Coltman, C.A.: J. Lab. Clin. Med., 74:967, 1969.

8. Rothenberg, S.P.: Blood 31:44, 1968.

9. Lau, K.S., Gottlieb, C., Wasserman, L.R. and Herbert, V.: Blood 26:202, 1965.

10. Hall, C.A. and Finkler, A.E.: Biochim. Biophys. Acta 78:234, 1963.

11. Hall, C.A. and Finkler, A.E.: J. Lab. Clin. Med., 65:459, 1965.

12. Hall, C.A. and Finkler, A.E.: Biochim. Biophys. Acta 147:186, 1967.

13. Lawrence, C.: Blood 33:899, 1969.

14. Bloomfield, F.J. and Scott, J.M.: Brit. J. Haemat., 22:33, 1972.

15. Allen, R.H. and Majerus, P.W.: J. Biol. Chem., 247:7702, 1972.

16. Allen, R.H. and Majerus, P.W.: J. Biol. Chem.: 247:7709, 1972.

17. Carmel, R. and Herbert, V.: Blood 40:542, 1972.

18. Herbert, V., Bloomfield, J., Stebbins, R. and Scott, J.: J. Clin. Invest., 52:39[a], 1973.

19. Bloomfield, J., Scott, J., Herbert, V. and Stebbins, R.: Fed. Proc., 32:892 (Abs.), 1973.

20. Taylor, K.B.: Lancet 2:106, 1959.

21. Schade, S.G., Abels, J. and Schilling, R.F.: J. Clin. Invest., 46:615, 1967.

22. Rothenberg, S.P., Kantha, K.R.K. and Ficarra, A.: J. Lab. Clin. Med., 77:476, 1971.

23. Rothenberg, S.P.: Nature 200:922, 1963.

24. Ghitis, J.: Am. J. Clin. Nutr., 20:1, 1967.

25. Kamen, B.A., and Caston, J.D.: J. Lab. Clin. Med., (In Press).

26. Rothenberg, S.P. and da Costa, M.: J. Clin. Invest., 50:719, 1971.

27. Retief, F.P. and Huskisson, J.: J. Clin. Path., 23:703, 1970.

28. McCall, M.S., White, J.D. and Frenkel, E.P.: Proc. Soc. Exp. Biol. Med., 134:536, 1970.

29. Ricker, R. and Stollar, B.D.: Biochem., 6:2001, 1967.

30. Rothenberg, S.P., Gizis, F. and Kamen, B.: J. Lab. Clin. Med., 74:662, 1969.

31. Rothenberg, S. P.: Proceeding of International Atomic Energy Symposium, "In Vitro" Procedures with Radioisotopes in Clinical Medicine and Research. 1970.

32. Rothenberg, S. P.: Anal. Biochem., 16:176, 1966.

33. da Costa, M. and Rothenberg, S. P.: Brit. J. Haemat., 21:121, 1971.

34. Waxman, S., Schreiber, C. and Herbert, V.: Blood 38:219, 1971.

35. *Rothenberg, S. P., da Costa, M. and Rosenberg, Z.: New Eng. J. Med., 286:1335, 1972.

36. da Costa, M. and Rothenberg, S. P.: J. Lab. Clin. Med., (In Press).

37. Waxman, S. and Schreiber, C.: Blood 42:281, 1973.

38. Rothenberg, S. P., da Costa, M., and Lawson, J.: Blood, (In Press).

*An erratum in this paper was corrected in a later issue of that journal, 287:208, 1972.

INTRODUCTION TO TOXICOLOGY

R. A. Scala, PhD

Esso Research and Engineering Company

Linden, New Jersey 07036

These few remarks are intended to set the stage for the three speakers who will follow. They will include some background on the history of toxicology, a review of the key concepts in this profession, and of the forces shaping investigations in toxicology and will conclude with a recitation of the steps leading to the development of a new chemical material. This generalized introduction will make it easier to follow the more specific and detailed presentations of Drs. Winek, Gibson and Carson.

It may be useful to begin with a definition of toxicologist and the profession of toxicology. A few moments spent with the Webster's Unabridged Dictionary will reveal that the words have their root in the Greek word, "toxon," meaning bow or arrow. From this have come the words "toxikon" meaning arrow poison, "toxicology," the science which treats poisons, effects, antidotes and their recognition, and finally, "toxicologist," one who practices that science. By collateral derivation, "toxology," and a number of related words have come into the language dealing with the sport of archery. The current definition of toxicology, promulgated by the Society of Toxicology a number of years ago, is "the

---

Abridged version of the presentation at a Symposium on Toxicology in Clinical Chemistry during the 1973 Meeting of the Eastern Analytical Symposium, November 15, 1973, Statler-Hilton Hotel, New York City.

quantitative study of the injurious effects of chemical and physical agents as observed in alteration of structure, function and response in living systems and includes the evaluation of safety."

One of the earliest toxicologists was Philippus Aurelous Paracelsus (1493-1541). He was a Swiss alchemist and physician and was also known as Theophrastus Bombastus VonHohenheim. It was he who articulated the concept that everything is poison if the dose is sufficient. Another important contributor to the development of the science was Mathieu Joseph Bonaventure Orfila who was born in Majorca in 1787 and died in Paris in 1853. He was a physician and chemist and was considered by many to be the father of modern medico-legal toxicology. He gathered information on what would today be called coroner's cases; conducted some research and from 1814 published a series of books on his finding. Orfila's major contribution was the unification by the middle 1800's, of toxicology by combining forensic and clinical elements along with analytical chemistry. This was shaken by a series of changes occurring from his time until now. The result is that the classical toxicology of Orfila is now only one area of the discipline, that is, the diagnosis and treatment of clinical poisonings and the associated analytical activities. One of the first departures was experimental pathology, closely associated with the work of Claud Bernard. More recently is a major expansion into safety evaluation. Most toxicologists devote their efforts to understanding the toxicity and making provisions for safe usage of industrial chemicals, drugs, food additives, cosmetics, household chemicals. Their evaluations are concerned not only with man but also with economic animals, domestic pets, wildlife, etc. Probably, the most recent development in the field of toxicology is a more rigorous study of plant and animal poisons and patterns of attack and defense. One might almost call this ecological toxicology. From that unified discipline of Orfila in 1850, there is now a highly fragmented discipline in which his forensic element comprises perhaps no more than 5%.

There are several major concepts with which a toxicologist is concerned. Toxicity, hazard and safety are three important, interrelated concepts. One should also consider dose, route, frequency, duration, species, age, sex, and site of action. These key words encompass the foundation for the presentations which will follow this. A comprehension of them is also required in order for the informed lay person to evaluate properly much of the toxicological news that is in the popular press. This news involves the effects of pesticides in the environment, the question of banning of materials for alleged or proven effects, the zero threshold limit for carcinogens in food, etc.

Toxicity is an expression of the adverse effects of a material. It might be considered a basic biological property of a compound similar to physical properties like boiling point and refractive index. The toxicity of a material describes the nature, degree and extent of undesirable effects. Hazard is an expression of the likelihood of this toxicity being manifest. It is quite clear then that materials may be hazardous but not very toxic or toxic but not very hazardous. Safety, in this context, is the provision of circumstances such as to minimize hazard.

One of the central concepts is that of dose. It is ordinarily expressed on a weight per unit body weight or surface area. But dose is frequently implied in expressions of concentration such as "milligrams per cubic meter of air" in the breathing space or "parts per million" in the diet, or "milligrams per liter" in drinking water or liquid, or "milliliters per square centimeter surface area of skin." Other measures of dose may involve peak or mean concentrations in blood; concentrations in target organs and concentrations in organs which are not target organs. Although dose is critical in terms of the amount of material delivered to the critical location, it is conveniently expressed as amount administered per unit weight or surface area of the individual involved. Since ordinarily the magnitude of the response is related to the magnitude of the dose, it is important to understand what dose means, how dose can vary, and how misleading dose administered or dose to which the animal is exposed can be in terms of understanding dose at the target organ. Closely associated with dose are the elements of frequency and duration. These may be determinants of total dose. They are particularly important in trying to determine whether a material is accumulated or readily excreted. This leads to such expressions of dose as body burden, biological half-life, and biotransformation. Finally, one must consider that certain combinations of frequency and duration may lead to constant doses (i.e. concentration x time = k). This concept was expressed in 1924 by Haber and is true for certain types of materials over relatively short intervals of concentrations and time.

Hayes (Proc. Royal Soc., Series B, 167:101, 1967) has shown how many of these factors are interrelated with respect to the toxicity of a number of pesticides. In his review of the published literature on pesticides toxicity, concentrating on acutely toxic effects, Hayes noted that for 67 materials tested by both the oral and dermal route, the ratio of the dermal LD_{50} to the oral LD_{50} averaged 4.2 but ranged from 0.2 to 21. In another series comparing the effect of sex, he found that for 65 pesticides, the oral LD_{50} for females was .9 times the LD_{50} for males with a range of .2 to almost 5. For 20 materials, the ratio of the oral LD_{50} in rats to the oral LD_{50} in other species averaged 1 but the range was 0.2 to 12. Not included in this calculation was norboramide, a material which is uniquely toxic to the rat, the ratio of rat LD_{50} to other animals being .0004. Hayes also made comparisons between the maximum no-effect level after 90 days of feeding and that after 2 years of feeding and the relationship of the age of the animal between newborn and several months of age to the acute oral toxicity. Probably the only generalization that can be developed from these data is that generalizations are extremely risky.

The dose-response curve is one expression of the relationship between dose magnitude and degree of response (Figure 1). For each region in the curve, an example will be given on a physiological basis and the analagous situation will be described for a toxic material. With low doses, effects fall in the homeostatic area (Region A). The physiological counterpart would be the minimal adjustments in size of blood vessel, heart rate and stroke volume associated with the change from a lying position to a standing position. For a foreign material breathed into the lung, there might be a change in the buffering of the blood as a response to a weak acid in order to keep blood pH constant. In the middle dosage range (Region B), one which represents normal response without a significant biological burden, the physiological analogy would involve the transition from standing to running. The compensatory processes are an increase in heart rate and an increase in minute volume of respiration. A comparable toxicological insult from inhaling an irritating foreign material might be slight airway constriction while still maintaining normal function. At higher doses, the limits of compensatory processes can be reached (Region C) and breakdown may occur. For the physiological analogy, the runner might collapse due to a stress on the cardiovascular system resulting from insufficient

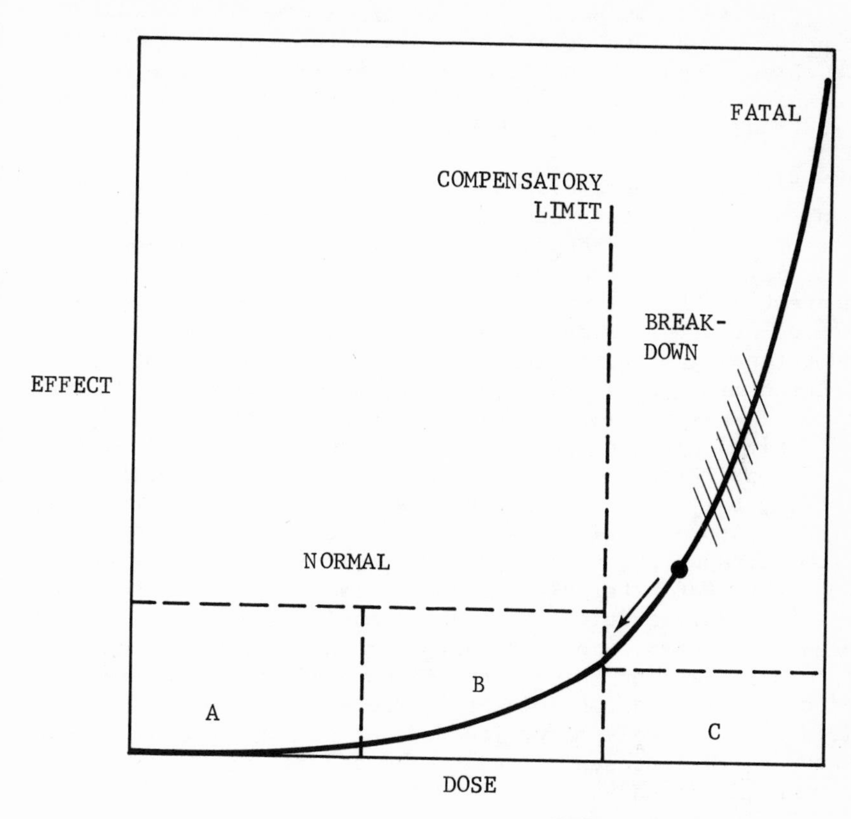

--FROM HATCH, AIHA J. 23:1, 1962

oxygen delivered or metabolic products taken away from the muscles. Following rest (a decreased function) full recovery would be expected. Toxicologically, this situation might arise with a severe chemical exposure resulting in bronchoconstriction and extremely or significantly decreased activity (i.e. hospitalization) followed by repair and recovery. At the highest doses, the obvious events can occur. The physiological or toxicological insult can exceed the repair potential of the organism and death or permanent disability may result.

Time does not permit further discussion of the dose-response curve with regard to the significance of whether the curve passes through the origin or not; the meaning of the shape of the curve; logarithmetic transformations of the curve, and discontinuity in the curve.

For an audience trained in scientific disciplines other than, although not toxicology, it may be useful to describe some of the forces shaping toxicological investigations. These may be viewed as five in number. First is the basic interest of the scientist himself. This involves incorporation of newer techniques and newer knowledge to increase understanding of toxicological processes. The advent of electron microscopy, scanning electron microscopy, and microchemical analysis for example, all provided new tools for the toxicologist to look at old problems. The availability of isotopes certainly facilitated pharmokinetic investigations, that is, the elucidation of pathways of absorption, distributions, storage, biotransformation and excretion. Gas chromatography has been a powerful tool in understanding precisely what materials have been associated with what effects. Parenthetically, mention may be made of a five year research program which the petroleum industry has sponsored at Carnegie-Mellon University on the toxicity of conventional petroleum hydrocarbon solvents. The earlier work in this field was done well before the advent of gas chromatographic analyses. The present studies will provide results well documented not only with regard to animal response but also with regard to the material studied.

A second force shaping toxicological investigations is heightened consumer interest. This interest has represented a form of pressure on otherwise intractable persons or institutions to do something about characterizing adverse effects of common materials. This consumer interest has sharpened responses of all toxicologists and made them aware of their responsibility to stand the scrutiny not only of their peers, but also of the public at large with regard to matters of safety evaluation.

This heightened consumer interest is related to the third force which may be described as general environmental concern. Any prudent individual or organization has an awareness of and responsibility for protection of the environment. It is easy to be eloquent about maintenance of the quality of life but it is now time to make quantitative statements and establish specific priorities. These are major public policy issues and biological scientists generally need to provide their reasoned assessment of effects and risks. Quantitative data are necessary to do this. Toxicologists and other biological scientists must then present their opinions and evaluations in order that legislators, government officials, and the public at large may be able to make informed judgments.

The fourth force is represented by regulatory developments. Toxicologists are currently concerned with several major statutes: The Food, Drug, and Cosmetic Act, Occupational Safety and Health Act, the Federal Environmental Pesticides Act, and finally the pending Toxic Substances Control Act. In each case, the legislature, in response to growing worker, consumer or public awareness and frequently sparked by some headlined event, has seen a fruitful area for activity. The resultant legislation has given many of the legislators high public visibility and increased the burden of the toxicologist, already laboring in a field where demands exceed resources. Reasoned scientific judgment in determining protocols for safety evaluation is being replaced by lists derived from regulations issued under the several laws. Now legislation protects the public from foods, drugs, and cosmetics, air and water, the chemicals in the workplace, toys, sleepwear and household products, and from agents used to control pests, predators and the crabgrass in the front yard. What few environmental materials have been omitted from this catalog will undoubtedly be covered by the pending Toxic Substances Control Act. This proliferation of laws resulted not only in inconsistencies but also in contradictions in regulations. Hopefully these will be resolved in the future in a reasonable fashion.

The analytical chemist has had a significant input in this area particularly with regard to what constitutes zero. Analytically, that value grows smaller as newer analytical techniques make it possible to find contaminants which had not been detected heretofore.

The final force shaping toxicological investigations is the general matter of product safety and support of product advertising. Many organizations have carried on toxicological investigations to support the safety of new products prior to introduction into the marketplace. These have been done before regulatory requirements and are today being done in the absence of specific regulatory requirements. Closely related to this is the conduct of scientific studies in order to support claims made in advertising or other labeling material. Simply reading the trade or popular press will uncover numerous examples of this.

To close, an example will be given of the manner in which a toxicologist puts the foregoing information together in a program of safety evaluation for a new product. Since other papers in this symposium will be concerned with food additives, drugs and forensic matters, a new cosmetic material will be used for illustration.

In any list of characteristics of the ideal cosmetic, safety would be included. The provision of absolute safety is absolutely impossible. However, a number of guidelines have issued and approximately three years ago the Pharmacology and Toxicology Committee of the Cosmetic Toiletry and Fragrance Association prepared such a set of suggestions (TGA Cosmetic Journal 2(4), Fall 1970). They listed the following basic tests as appropriate for most products: epidermal (for irritancy, toxicity and sensitization); ophthalmic irritancy; and oral administration for acute toxicity. In addition to these basic tests, the Committee felt that there were special tests necessary for certain types of products. For example, products applied to the skin where there is exposure to sunlight should be subjected to photosensitization studies in humans. Lipsticks, dentifrices, mouthwashes, bath preparations and personal hygiene products which may involve mucous membrane exposures should be subjected to irritancy tests in animals and controlled use tests in humans. Aerosols, sprays, face powders, talc, etc. should be subjected to inhalation tests in animals and finally, lipsticks, dentifrices, mouthwashes, and certain raw materials should be subjected to feeding studies in animals because of the possibility of accidental or deliberate ingestion by humans. The list is not meant to be all inclusive but simply points out the range of testing which needs to be done. With this, it is still possible to find areas where problems can arise in actual usage which were not anticipated in this development program. Therefore, the general program should be preceded by literature review and followed by controlled marketing of the new material where adverse reactions could be quickly identified before general marketing was attempted. Even then, the possibility of sensitization or idiosyncratic responses cannot be ruled out because the number of subjects involved is too small to pick up, except accidentally, these highly uncommon responses. Similarly most experimental studies cannot anticipate every possible manifestation of misuse. And finally, there is constant need for reexamination of materials as methodologies improve with time.

This introductory talk has been a short review of the background of toxicology, some of the key concepts, some of the kinds of things in which toxicologists find themselves involved and the reasons for their involvement, and a few of the steps in the safety evaluation of a new material.

FORENSIC TOXICOLOGY

Charles L. Winek, Ph.D.

Chief Toxicologist, Allegheny County Coroner's Office and Prof. Toxicology, Duqesne Univ.

Pittsburgh, Pennsylvania 15219

Toxicology has become almost a household word. Stimulated by man's concern with health in general and the problems associated with staying healthy, toxicology has become a very popular and paramount science in the last decade. Air and water pollution, drugs and alcohol, carcinogens and many other problems are responsible for the prominence of toxicology.

Despite the popularity of toxicology, it continues to be mispronounced and misspelled. Table 1 contains a listing of the many titles that have been given to the author on letters addressed to him at the University and at the Coroner's Office. The ultimate was the receipt of a letter addressed to the "Chief Taxidermist," Allegheny County Coroner's Office!!!

TABLE 1 - - Misspelled titles for TOXICOLOGIST

Toxirologist	Toxiologist
Coxicologist	Toxologist
Toxsologist	Toxeologist
Tarxologist	Troxologist
Troxicologist	Taxiologist
Taxologist	Tixicologist

Briefly defined, toxicology is the science of poisons. In most areas of toxicology, the effect of a poison is measured in some manner. For example, in clinical toxicology the severity of symptoms are determined and generally the amount of toxicant that gained entry into the body is quantitated. In forensic toxicology, the amount of poison present in the body tissues after death is quantitated and the significance of the concentration level is interpreted. For example, in a recent case reported by Tucker (1), a 17-year old female ate an unknown quantity of PMA (p-methoxy-amphetamine) and died some six hours later. The blood level of PMA was found to be 0.04 mg%, with the heart and kidney containing about 1.0 mg%. These findings, in the absence of any other cause of death, give an indication of the amount of drug necessary to cause death. The accumulation of data of this type from death investigations is useful to other toxicologists in arriving at the correct diagnosis of death. Admittedly, publication of this type of information is archival in nature, but extremely important from a medical-legal point of view. The importance of measuring the amount of drug, chemical, or poison that gains entry into a living organism is important in most areas of toxicology whether it be clinical,forensic, or research. This implies that analytical toxicology is basic to toxicology and, further, that toxicologists must be adequately trained in all aspects of modern analytical chemistry. Information on drug and chemical blood levels is presented later.

Many interesting and heavily publicized cases involving drugs, chemicals, poisons and toxicants have served as stimuli to the general awareness of toxicology. These cases constitute some of the recent history of toxicology and include the following:

1. The Thalidomide episode.
2. The concern with DDT problems and pesticides in general.
3. Fluoridation of drinking water.
4. Cyclamates and saccharin.
5. The Coppolino case (Succinylcholine).
6. The pill (birth control pills).
7. Hexachlorophene (percutaneous absorption).
8. Methadone and heroin addiction.
9. Mercury in water and its inclusion in edible fish.
10. Lead pollution.

11. Carbon monoxide from certain autos and CO pollution of the air in general.
12. The drug abuse problem and emphasis on urinalysis for drugs.
13. Drug deaths.

These items have all contributed to the national awareness of toxicity, toxicology, and the toxicologist. The majority of these items are related to chronic toxicity as opposed to acute toxicity.

If we consider the various roles that toxicology plays in the total medical, legal, and health areas, a toxicologist is involved with human life from the time of conception (and even before), to the time of death, and even after burial in certain cases where exhumation is required.

A toxicologist is not a limited or inhibited scientist in the same sense that a urologist or obstetrician works almost exclusively below the waist. A toxicologist has, in general, a knowledge of the specialty and then may elect to limit himself within the specialty because of economic, political, or personal reasons.

A forensic toxicologist is a scientist with major interest in chemical analyses of materials (physical and biological) and interpretation of his findings. He is generally located in a coroner's office, medical examiner's office, health department, police laboratory, hospital laboratory, or private service laboratory, depending on the specific state, county, or local laws and finances. He is not just the local expert that the district attorney calls upon to testify when he needs him or wants him. Admittedly, he is generally concerned with acute toxicity, though not always, and is called upon to testify in criminal cases, generally by the prosecution. He can and does become involved, if only in a consulting capacity, in criminal cases for the defense and civil cases as well.

The Specific Roles

The forensic toxicologist has several roles in the criminal justice system as well as community services. These roles would be classified under the following general headings:

1. Investigative (on scene).
2. Preservation of evidence (chain of evidence).
3. Analyses of materials.
4. Reporting and recording of results.
5. Interpretation of toxicological data.
6. Consultation (pathologist, police, attorney).
7. Expression of opinion (oral and written).
8. Expert testimony.
9. Research (methodology, improvement of analytical techniques and procedures).
10. Education.
11. Clinical Toxicology Service.
12. Drug Identification.
13. Poison Consultation.

The Forensic Toxicologist

Before proceeding to a brief consideration of some of these roles, some discussion of the training and qualifications of a toxicologist is merited. Unlike obtaining a degree in medicine and then following this with internship and residency in a specialized area -- with the end result of a man being board certified in forensic pathology -- a toxicologist may have a B.S., or Ph.D. degree in any number of areas ranging from various specialties in chemistry to pharmacy, pharmacology, pharmaceutics, and more recently, even toxicology. The remainder of his training comes from the less formal "on the job" or apprentice type training. This is not to say that the university course work or research work that he did is unrelated to toxicology. In most cases, the university experience and training relates well to the training of a toxicologist. If a degree program existed at a university, many of the courses already offered in the department could be applied toward a degree in toxicology.

It should be pointed out that beginning in the early 1960's, degree programs leading to the Master's and Doctor's Degree in Toxicology were begun, and are in existence today. One such program exists at Duquesne University School of Pharmacy.[2] Prior to that time, most toxicologists were toxicologists by official title only, and not because they were awarded an academic degree in toxicology. This does not mean that they are unqualified; on the contrary, they are more qualified because of the great number of experience. I should like to caution, however, about the individual who presents himself as a toxicologist and in actuality, by training or experience, is not. Example: A young man who has a State title of Chemist I, testifying in a case involving alcohol. In qualifying this individual, he indicates that he has about 4-1/2 years experience. The defense attorney, on examination, asks about this experience, and learns that he went to school for 4 years, has a B.A. in Biology, has never had a chemistry course, and has been working in the laboratory for about 4 months.

The qualifications of forensic chemists and drug testing procedures were reported on by Stein _et al_ (3). They indicate in general that many forensic chemists performing drug analyses are not properly trained in some of the basic and specialized areas needed to identify adequately an unknown illicit drug sample with a reasonable degree of scientific certainty. They cite several cases in which the forensic chemist was not qualified by degree or chemical background, but served for years as a "forensic chemist" by official title. I personally believe that persons desiring to change jobs and gain employment in the area of analytical, forensic chemistry or toxicology, should read Stein's article and investigate in detail the magnitude of the job and the demands placed on such a job. This article should be digested by those considering forensic toxicology or chemistry as a possible career. Fochtman and Winek (4) pointed out the problem involved with the proper identification of marihuana using the Duquenois-Levine color test. Various brands of coffee give a false positive test for marihuana and if the Duquenois-Levine test is used as the only means of identification, a possible miscarriage of justice could result. Ego bias and improper testing procedures can lead to false chemical evidence and a miscarriage of justice. There is no excuse for the generation of faulty scientific data with the existence of our modern analytical techniques and excellent training programs at both the graduate and undergraduate levels.

It should be stressed that a forensic toxicologist is qualified generally because of his training, research, and experience, all of which take time. A trial attorney has a degree, but it is his experience in court that makes him a good trial attorney. He gains this experience by working in a firm that deals with criminal law. A toxicologist has a degree, but it is his training and experience in a forensic toxicology laboratory that makes him a good forensic toxicologist.

Research

The forensic toxicologist must be concerned with improving methods of analyses for many chemicals and drugs. He must occasionally develop a method of analysis for a new drug or a metabolite of a drug. He must be aware of the differences in accuracy and sensitivity of one method as opposed to another. Newer instrumental methods of analyses such as mass spectrometry can increase the detection limits for drugs and increase the speed of analyses. However, many governmental offices can be held up in improving the methods and procedures because of financial limitations. Lawyers, particularly, should keep in mind that politicians are not about to allocate funds for the dead (they can't vote) when monies are needed elsewhere for programs. Research on post mortem chemical changes and concentrations of drugs and chemicals is limited because of funding. What happens to the carbon monoxide concentration in tissue after a body is embalmed and buried for 4 years? This question and others like it may go unanswered for years because research on the subject is lacking. The answer may be important to specific legal proceedings, but obtaining funds for research in this area is difficult.

The forensic toxicologist must keep abreast of newer methodology and instrumentation and thereby keep his testing procedures current. Certain instruments are a must for a properly functioning forensic toxicology laboratory. In addition to the usual and expected basic laboratory glassware and hardware, the following are considered absolute:

1. Gas Chromatograph (I prefer Perkin Elmer Model 3920 and consider 3 units the minimum for a busy lab).
2. Spectrophotometer (I prefer the Bausch & Lomb 505 and consider 2 the minimum.
3. Infrared spectrophotometer.
4. Spectroflurophotometer (Perkin Elmer Model MPF-2A
5. Colorimeter (one is adequate).
6. Atomic Absorption Spectrophotometer (with various lamps for common metals).

There are also several specialized items that are needed such as Conway diffusion cells and Burrell wrist action shakers. In addition to a fume hood, a perchloric acid resistent hood or one with a water curtain is desirable. Never waste money on "toys" or unproven gadgets. When buying an expensive piece of equipment or some new "time - saver" or "personnel eliminator," insist on a free two to three week trial period to see how it works for you when the salesman is gone!

Education

The toxicologist has the obligation to provide education on the general subject of forensic toxicology for his professional colleagues and for the general populace in his area. He serves as the expert for local problems dealing with the environment and will find himself on various committees involving lead poisoning, drug abuse, pesticides, alcohol, sports medicine, poison control, and consumer protection. He, in general, can serve his community with toxicity information, and in this way he educates others.

The forensic toxicologist can provide continuing education for police, lawyers, and physicians and can, as most are, be involved with university graduate programs in toxicology.

On Scene Investigation

There are many cases in which the toxicologist should perform "on scene" investigations for the purpose of discovering the cause or source of a toxicant involved in a death. Because Pittsburgh is heavily industrialized, we have the occasion to do "on scene" investigations involving deaths that may be related to toxic gases, fumes, solvents, etc. This type of investigation by a toxicologist may lead to discovery of criminal negligence. On-scene investigation of specific cases certainly lends strength to the testimony given by a toxicologist in a legal proceeding, although admittedly, it is not necessary in every fatal auto accident, suicide, or homicide.

We were investigating an abortion death, and received a call from the police to accompany them to a suspect's house where an abortion was in progress. This proved most useful in the subsequent trial because of my personal findings of drugs, "secret abortion oil," surgical equipment, and the abortionist's statements to me.

Preservation of Evidence

The forensic toxicologist is responsible for the proper handling of all evidence of concern to his area in a given case. Proper chain of evidence should be documented with receipts or records indicating the persons responsible for the transfer of materials to be analyzed by him. For example, many drunken driving cases are won or lost, depending upon your point of view, because no one can produce the person who drew the blood sample or can show that the sample belonged to the person being charged with the violation. The analysis of the blood sample is a complete waste of the toxicologist's time in these cases because of poor chain of evidence. If the toxicologist exerts himself and insists on proper documentation of chain of evidence, from the time of blood withdrawal to delivery to his laboratory, before the analysis is performed, a more productive result is obtained. Forms for use in chain of evidence are shown in Figs. 1 and 2. It is, however, still amazing to me that many defense attorneys stipulate to the toxicologist's report and accept it without question.

Analyses of Materials

A major job of the forensic toxicologist is to identify toxicants (chemicals, drugs, poisons), in biological material. This identification is carried out using accepted analytical methods and modern instrumentation which vary from one laboratory to another.

In many cases today, simple qualitative analysis is not enough for legal purposes. It may be sufficient for death certificate purposes, but for legal purposes, quantitation of the chemical is necessary, i.e., the tissue concentration and/or blood level of the chemical must be determined.

I think that a major role for the forensic toxicologist begins with the determination of the concentration of

HOSPITAL NAME

PATIENT'S NAME______________________________

PATIENT'S ADDRESS____________________________

____

DATE: ______________________________

Area was swabbed with aqueous Zephiran. No Alcohol was used to swab the area before the blood sample was withdrawn.

TIME OF WITHDRAWAL____________________________

BLOOD DRAWN BY______________________________

WITNESSED BY______________________________

____

____

POLICE DEPT.______________________________

PHYSICIAN INVOLVED____________________________

Code number for tube containing blood,
(only if used).____________________

FIG. 1 Hospital form for documentation of blood sample.

ALLEGHENY COUNTY CORONER'S OFFICE

542 Fourth Avenue

Pittsburgh, Pennsylvania

DR. C. L. WINEK
Chief Toxicologist

RECEIPT FOR TOXICOLOGY DEPT.

NAME OF PATIENT____________________

Name of person delivering specimen(s)____________________

Hospital or Police Department Name____________________

Time and Date of Delivery____________________

COMMENTS: (if any)____________________

Received by____________________

FIG. 2 Receipt form for chain of evidence

chemical found. Here he must render an interpretation of the results; his opinion of the significance of a given concentration of antihistamine in the brain tissue of an airplane pilot following a fatal crash, his opinion of the amount of benzene found in the brain of a deceased industrial worker, his opinion of the concentration of alcohol in the blood of a victim that received 14 pints of blood before death, his opinion of the concentration of lead or carbon monoxide, or acetone, or benzene found in the tissues of deceased workers. This interpretation of toxicological findings, either his own or in reviewing the report of another toxicologist, is a unique and special area for the forensic toxicologist.

The subject of drug and chemical blood levels has been described in several previous publications (5,6,7,8,9,10,11). Readers interested in the subject are referred to these articles. Blood level or concentration of a drug indicates the amount of drug present in the blood at the time of analysis. The amount present can reflect the seriousness of a poisoning in a living patient, degree of exposure of an industrial worker to a volatile toxicant, etc. Blood levels have been divided into therapeutic, toxic, and lethal, and have been defined as follows:

Therapeutic blood level. Fochtman and Winek (12) defined a therapeutic blood level as that concentration of drug present in the blood (its serum or plasma) following therapeutically effective dosage in humans.

Toxic blood level. The concentration of drug or chemical present in the blood (its serum or plasma) that is associated with serious toxic symptoms in humans.

Lethal blood level. The concentration of drug or chemical present in the blood (its serum or plasma) that has been reported to cause death, or is so far above reported therapeutic or toxic concentrations that one can judge that it might cause death, in humans.

Drug and chemical blood levels are given in Tables 2, 3, and 4. These data are offered as a guide for the chemical, analytical, and forensic toxicologists as well as physicians and lawyers. A blood level range is indicated in the Tables where the information is available. In some instances a single value is given because additional information or experience is lacking.

TABLE 2 DRUG BLOOD LEVELS

DRUG	THERAPEUTIC	TOXIC	LETHAL
Acetaminophen (Tylenol)	1 to 2 mg%	40 mg%	150 mg%
Acetohexamide (Dymelor)	2.1 to 5.6 mg%	-	-
Aminophylline	2 to 10 mg%	-	-
Amitriptyline (Elavil)	-	40 mcg%	1.0 to 2.0 mg%
Amphetamine	-	-	0.2 mg%
Barbiturates			
Short acting	0.1 mg%	0.7 mg%	1 mg%
Intermediate acting	0.1 to 0.5 mg%	1 to 3 mg%	3 mg% & >
Phenobarbital	ca.* 1.0 mg%	4 to 6 mg%	8 to 15 mg% & >
Barbital	ca. 1.0 mg%	6 to 8 mg%	10 mg% & >
Bromide	5.0 mg%	50 mg%	200 mg%
Chloral hydrate (Noctec)	1.0 mg%	10 mg%	25 mg%
Chlordiazepoxide (Librium)	0.1 to 0.3 mg%	0.55 mg%	2 mg%
Chlorpheniramine	-	2 to 3 mg%	-
Chlorpromazine (Thorazine)	0.05 mg%	0.1 to 0.2 mg%	0.3 to 1.2 mg%
Chlorpropamide (Diabinese)	3.0 to 14.0 mg%	-	-

TABLE 2 (continued) DRUG BLOOD LEVELS

DRUG	THERAPEUTIC	TOXIC	LETHAL
Chlorprothixine (Taractan)	0.004 to 0.03 mg%	--	--
Desipramine (Norpramin)	.059 to 0.14 mg%	--	0.3 mg%
Dextroamphetamine	--	--	0.2 mg%
Dextropropoxyphene (Darvon +)	5 to 20 mcg%	--	5.7 mg%
Diazepam (Valium)	0.05 to 0.25 mg%	0.5 to 2.0 mg%	2.0 mg%
Diphenhydramine (Benadryl)	0.5 mg%	1 mg%	--
Diphenylhydantoin (Dilantin)	0.5 to 2.2 mg%	$>$ 5 mg%	10 mg%
Ethchlorvynol (Placidyl)	ca. 0.5 mg%	2 mg%	15 mg%
Ethyl ether	90 to 100 mg%	--	140 to 180 mg%
Glutethimide (Doriden)	0.02 mg%	1 to 8 mg%	3 to 10 mg%
Hydromorphone	--	--	10 to 30 mcg%
Hydroxyzine (Atarax) (Vistaril)	0.5 mg%	ca. 1.0 mg%	--
Imipramine (Tofranil)	0.2 to 0.6 mg%	--	--

TABLE 2 (continued) DRUG BLOOD LEVELS

DRUG	THERAPEUTIC	TOXIC	LETHAL
Meperidine (Demerol)	60 to 65 mcg%	0.5 mg%	ca. 3 mg%
Meprobamate	1 mg%	10 mg%	20 mg%
Methadone	48 to 86 mcg%	--	--
Methamphetamine	--	--	4 mg%
Methapyrilene	--	3 to 5 mg%	--
Methaqualone	0.5 mg%	1 to 3 mg%	3 mg% & >
Methylenedioxyamphetamine (MDA)	--	--	0.4 to 1.0 mg%
Methyprylon (Noludar)	1.0 mg%	3 to 6 mg%	10 mg%
Morphine	--	--	0.005 mg% (free morphine)
Nitrofurantoin (Furadantin)	0.18 mg%	--	--
Nortriptyline (Aventyl)	0.1 mg%	0.5 mg%	--
Oxazepam (Serax)	0.1 to 0.2 mg%	--	--
Paraldehyde	ca. 5.0 mg%	20 to 40 mg%	50 mg%
Pentazocine (Talwin)	0.014 to 0.016 mg%	--	--

TABLE 2 (continued) DRUG BLOOD LEVELS

DRUG	THERAPEUTIC	TOXIC	LETHAL
Phenylbutazone (Butazolidin)	ca. 10 mg%	--	--
Probenecid (Benemid)	10 to 20 mg%	--	--
Propranolol (Inderal)	0.0025 to 0.02 mg%	--	--
Quinidine	0.3 to 0.6 mg%	--	--
Quinine	--	--	1.2 mg%
Salicylate (acetylsalicylic acid)	2 to 10 mg%	15 to 30 mg%	50 mg%
Sulfadimethoxine (Madribon)	8 to 10 mg%	--	--
Sulfisoxazole (Gantrisin)	9 to 10 mg%	--	--
Thioridazine (Mellaril)	0.10 to 0.15 mg%	1.0 mg%	--
Tolbutamide (Orinase)	5.3 to 9.6 mg%	--	--
Trimethobenzamide (Tigan)	0.1 to 0.2 mg%	--	--
Zoxazolamine (Flexin)	0.3 to 1.2 mg%	--	--

*ca is from the Latin circa, meaning approximately

+Liver levels are about twenty times as high as blood levels

TABLE 3 CHEMICAL BLOOD LEVELS

CHEMICAL	NORMAL*	TOXIC	LETHAL
Acetone	--	20 to 30 mg%	55 mg% (P)+
Benzene	--	any measurable	0.094 mg% (P)
Carbon monoxide	--	15 to 35%	50%
Carbon tetrachloride	--	2 to 5 mg%	--
Chloroform	--	7 to 25 mg%	39 mg%
DDT	0.013 ppm	--	--
Dieldrin	0.0014 ppm	--	--
Dinitro-o-Cresol	--	3 to 4 mg%	7.5 mg%
Ethanol	--	0.15%	0.35% & >
Ethylene glycol	--	0.15%	21.0 mg% (P)
Oxalate	0.2 mg%	--	1.0 mg%
Hydrogen sulfide	--	--	0.092 mg% (P)
Isopropanol	--	340 mg%	--
Methanol	--	20 mg%	80 mg%
Methylene chloride	--	--	28 mg% (P)
Nicotine	--	1 mg%	0.5 to 5.2 mg%
Toluene	--	--	1.0 mg% (P)
Tribromoethanol	--	--	9 mg%
Trichloroethane	--	--	10 to 100 mg%

*Accepted as being normal or expected because of our environment
+Levels indicated with a (P) are cases investigated by the author

TABLE 4 BLOOD LEVELS OF METALS, METALLOIDS, AND INORGANIC SUBSTANCES

CHEMICAL	NORMAL*	TOXIC	LETHAL
Aluminum	0.013 mg%	--	--
Arsenic	0.0 to 0.002 mg%	0.1 mg% (P)+	1.5 mg% (P)
Boron (Boric acid)	0.08 mg%	4 mg%	5 mg%
Beryllium	tissue levels generally used (lung & lymph)	--	--
Cadmium	--	0.005 mg%	--
Copper	40 mcg%	0.54 mcg%	--
Cyanide	0.015 mg%	--	0.5 mg% &>
Fluoride	0 to 0.05 mg%	--	0.2 mg%
Iodide	--	--	100 mg%
Iron	50 mg% (RBC)	0.6 mg% (serum)	--
Lead	0.005 to 0.13 mg%	0.07 mg%	--
Lithium	0.6 to 1.2 mEq/liter	2.0 mEq/liter	2.0 to 5.0 mEq/liter
Magnesium			5 mg%
Manganese	0.015 mg%	0.46 mg%	--
Mercury	0.006 to 0.012 mg%	--	--
Nickel	0.041 mg%	--	--
Phosphorus	tissue levels generally used	--	--
Thallium	--	--	--
Tin	0.012 mg%	--	--
Zinc	0.9 mg%	--	--

*Accepted as being normal or expected because of our environment.
+Levels indicated with a (P) are cases investigated by the author.

Blood level data is useful medically and/or legally in the following:

1. To determine the adequacy of treatment in poisoning cases.
2. To establish the generic equivalency of drugs.
3. To determine the cause of death when a drug or chemical is detected in the blood.
4. To determine the therapeutic effectiveness of a drug.
5. To determine the extent of exposure to an industrial or environmental toxicant.

Obviously there are factors that can affect the blood level value. As indicated several times above, interpretation of analytical results is an important aspect of the toxicologist's results. All of the following factors should be considered when interpreting a blood level:

1. Combinations of drugs and/or chemicals.
2. Time between ingestion and sampling of blood (time for absorption to take place).
3. Tissue binding of drug or chemical.
4. Method of analysis.
5. Dosage of drug or amount of exposure to a chemical.
6. Pathological state of the patient.
7. Time of analysis of blood sample (loss of chemical due to storage).
8. Biologic half-life of drug and dosage regimen (accumulation, or tissue storage).
9. Age, sex, race, and weight of the patient.
10. Time of discovery of a poisoned patient or death victim (absorption and diffusion factors).

Reporting and Recording of Results

Proper documentation of the laboratory findings is essential to a properly operated forensic toxicology laboratory. The laboratory work report should be filed after a typed copy has been supplied to the proper parties. In many laboratories, a lab record book is also kept. In cases where a report is "lost," there is still available a permanent lab book, as well as the lab work report. Reports should indicate date of analyses, as well as by whom they were done, particularly in larger facilities where there may be a dozen or more toxicologists working in the same lab. It is also desirable to note the time of analysis and also confirmation of positive results by another toxicologist in the same lab. This affords a type of continual quality control.

Consultation and Opinion

The forensic toxicologist may consult with the pathologist or police in a specific case currently being investigated. It may consist of a simple answer to a simple question, "How much liver do you need for morphine analysis," or "Is 0.18 percent blood alcohol legally drunk?"

Whether or not a forensic toxicologist is consulted by his colleagues, reflects their personal opinion of him. I guess this is called evaluation, or his reputation. If the forensic toxicologist develops a reputation of "being good on anesthetic deaths," he generally receives consultations involving such cases. Consultation with a forensic toxicologist should be considered, even if you only want to know if there is a better way to conduct a specific test, or if you want to be prepared to cross-examine a forensic toxicologist in a case. I recall a case where I was on the witness stand and another forensic toxicologist was feeding handwritten questions to the defense attorney. He did not call the toxicologist as a witness, but simply held on-the-spot consultations with him in court.

Consultation with a forensic toxicologist is extremely important in a criminal case involving the report of analyses and opinion of another forensic toxicologist. First of all, you learn whether or not the findings and opinion are generally acceptable, or if they are controversial, or completely or partially unacceptable. A trial attorney learns how to cross-examine the forensic toxicologist, using the proper type of questioning.

I testified in a case involving a man charged with abortion death. He was alleged to have supplied ergot to a woman who died from taking ergot in attempting abortion. A chemist had testified that he identified ergot in her stomach contents. His report said that a substance similar to ergot was present in the stomach contents.

After reviewing his analytical procedure, I testified that he could not possibly identify ergot by the procedure he used.

BIBLIOGRAPHY

1. Tucker, R., Bull. Intern Assoc. For. Tox. 9, #3 & 4, p. 15, 1973.

2. Winek, C. L. and Ruggiero, J.S. A Training Program in Toxicology. Am. J. Pharm. Ed., 31, Nov. 1967.

3. Stein, B., Laessig, R.H., and Indriksons, A., Wisconsin Law Review, Vol. 1973, No. 3, 727 - 789, 1973.

4. Fochtman, F.W. and Winek, C.L., Clinical Toxicology, Vol. 4, 287, 1971.

5. Winek, C.L., in Legal Medicine Annual, C. H. Wecht, ed. Appleton-Century-Crofts, 67-77, 1970.

6. Winek, C.L., Clinical Toxicolgoy, Vol. 3, No. 4, 541-549, 1970.

7. Winek, C.L., Amer. J. Hosp. Pharm. Vol. 28, 351-356, 1971.

8. Winek, C.L., in Medical Pharmacology, A. Goth, 6th ed., The C.V. Mosby Co., 1972.

9. Winek, C.L., in Narcotics and Narcotics Addiction, Maurer & Vogel, Charles C. Thomas publisher, 1973.

10. Winek, C.L., in Modern Drug Encyclopedia. The Yorke Medical Group, 12th ed. 1973.

11. Winek, C.L., in Legal Medicine Annual, C. H. Wecht, ed., Appleton-Century-Crofts, 115-120, 1973.

12. Fochtman, F.W. and Winek, C.L. J. For Sci. Vol. 14, 213-218, 1969.

ROLE OF BIOTRANSFORMATION REACTIONS IN PRENATAL AND POST-NATAL CHEMICAL TOXICITY

James E. Gibson

Department of Pharmacology
Michigan State University
East Lansing, Michigan 48824

Drugs and other chemicals may produce qualitatively and quantitatively different toxicologic effects in embryos, fetuses and newborns than in the adult of the same species. When such toxicity occurs during embryogenesis congenital malformations may result. On the other hand, agents which produce toxicity during late fetal development, or during the neonatal period, may impair development or produce toxicity qualitatively similar to that seen in the adult. During embryogenesis there are embryological and non-embryological factors which determine an organism's susceptibility to an agent, such as the nature of the agent, the dose, and maternal physiology (Wilson, 1965). Agents which disrupt embryonic development may initially act at any of the processes essential for normal development, e.g., cell division, differential gene expression, changes in the surface properties and shapes of cells, or general cellular metabolism. Most organs have a period of particular susceptibility to teratogens which probably coincides with the early and critical developmental events for that organ. During the fetal and neonatal period, however, after differentiation of cells is largely completed, chemical toxicity may involve interference with cell growth.

Exposure of drugs or chemicals to the developing embryo, fetus or neonate may produce adverse effects on these systems as a result of a particular sensitivity to the agent or as a result of alterations in metabolic fate and distribution which influence the relative activity of the drug or chemical as compared with adults. Thus, the toxic manifestations of a drug or chemical in the developing mammal may be the result of either of these phenomena separately or in combination.

Many developmental pharmacology studies have attributed the failure of drug efficacy in young animals to be secondary to a low conversion of the parent drug to an active metabolite or, conversely, the increase in drug or chemical toxicity secondary to a failure of the immature organism to detoxify the parent drug or chemical. The system responsible for many metabolic conversions of drugs and chemicals _in vivo_ may be called the "toxication-detoxication system" (Fouts, 1972) and may use as substrates drugs, foreign chemicals, endogenous steroids and nonlipids, fatty acids, or vitamins. The metabolic changes in drugs or chemicals may involve oxidation or reduction, synthetic or hydrolytic reactions, and may be carried on in many organ systems. Ordinarily a chemical substance taken into the body will be converted enzymatically by the "toxication-detoxication system" to a more polar, more water soluble metabolite that will be readily excreted. The system, therefore, may operate to terminate the actions of chemical substances in the body. Alternatively, these biotransformations may serve to convert a biologically inactive compound into one having therapeutic effectiveness and/or toxicity. Typically, these reactions are thought of as occurring primarily in the endoplasmic reticulum of hepatic cells (Conney, 1967), although other organs may also participate in these metabolic processes and indeed fractions of the cell besides the endoplasmic reticulum may be involved.

Many studies have demonstrated that the "toxication-detoxication system" responsible for the metabolism of drugs and other substances was either absent or at barely detectable levels in the fetus and newborn of many animal species (Fouts and Adamson, 1959; Jondorf _et al_., 1959; Soyka, 1969; Wilson, 1970; Short and Davis, 1970; Henderson, 1971;

Mac Leod *et al*., 1972; Fouts and Devereux, 1972). Adult levels of activity were reached at varying times depending on the individual enzyme pathways and the animal species studied. Thus, the consequences of low drug or chemical metabolizing activity in the fetus and newborn might be predicted to influence the action of a drug or chemical during those stages of development. A recent tabulation of LD_{50} values in newborn and adult animals has emphasized this principle (Goldenthal, 1971).

CYCLOPHOSPHAMIDE: AN AGENT THAT AFFECTS PERINATAL DEVELOPMENT

Cylcophosphamide [1-bis(2-chloroethyl)-amino-1-oxo-2-aza-5-oxa-phosphoridin] is an antineoplastic agent which may exert cytotoxic effects by alkylation of cellular macromolecules. Cyclophosphamide is inactive as a cytotoxic agent *in vitro* but is highly cytotoxic *in vivo*. The *in vivo* activation of cyclophosphamide occurs by means of the hepatic "toxication-detoxication system" mentioned above (Brock and Hohorst, 1963, 1967; Cohen and Jao, 1970; Connors *et al*., 1970; Sladek, 1971). This bioactivation is inhibited by SKF-525A (2-diethylaminoethyl-2,2-diphenylvalerate hydrochloride) pretreatment (Brock and Hohorst, 1967; Field *et al*., 1972; Gibson and Becker, 1968a) and stimulated by phenobarbital pretreatment (Rauen and Kramer, 1964; Gibson and Becker, 1968a; Field *et al*., 1972).

Hill *et al*. (1972) proposed a pathway for the metabolic activation of cyclophosphamide in mouse liver. The carbon atom adjacent to the ring nitrogen was proposed to be oxidized by the "toxication-detoxication system" to form 4-hydroxycyclophosphamide and then was subsequently oxidized to aldophosphamide (Figure 1). The formation of carboxyphosphamide, the oxidation product of aldophosphamide, was catalyzed by a soluble enzyme having the properties of aldehyde oxidase. Aldophosphamide inhibited clone formation of human epidermoid carcinoma No. 2 cells and produced toxicity in L1210 cells and was considered to be the cytotoxic metabolite. In contrast, carboxyphosphamide produced little cytotoxicity. Thus, the biological activity of cyclophosphamide is thought to be due to its metabolic activation to a highly electrophilic substance which will alkylate nucleophilic groups, many of which are important biological moieties such as sulfhydryl groups, carboxyl and amino groups. The primary biochemical lesion appears to be an inhibition of DNA, but not RNA or protein synthesis (Roberts et al., 1968; Wheeler and Alexander, 1969; Short et al., 1972; Short and Gibson, 1973).

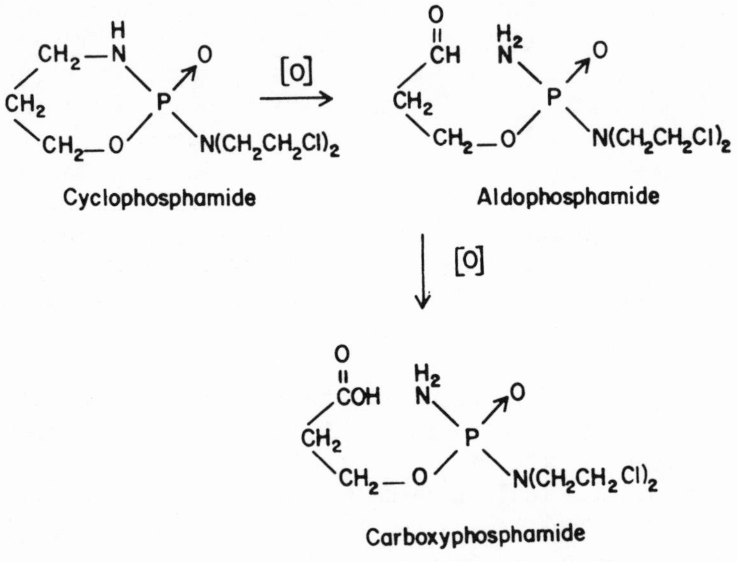

Figure 1. The metabolism of cyclophosphamide (Hill et al., 1972) proceeds via hepatic oxidation to aldophosphamide, the agent considered to be the cytotoxic metabolite. The formation of carboxyphosphamide, the oxidation product of aldophosphamide, is catalyzed by a soluble enzyme having the properties of aldehyde oxidase.

CYCLOPHOSPHAMIDE PRENATAL TOXICITY

Cyclophosphamide is teratogenic in a variety of mammalian species including rabbits (Gerlinger, 1964), rats (Murphy, 1962; Wilson, 1964; Kreybig, 1965), and mice (Hackenberger and Kreybig, 1965; Shoji and Ohzu, 1965; Gibson and Becker, 1968b). The teratogenicity of cyclophosphamide was similar to all species. In mice, single doses administered on days 10-15 of gestation produced a variety of teratogenic effects, a reduction in fetal growth, and an increase in lethality. The anomalies produced included gross defects (cleft-palate, exencephaly, digital defects and kinky tail), skeletal anomalies (polydactyly, syndactyly, ectrodactyly, fusion of the long bones, curvature of the long bones, and missing ribs), and soft tissue malformations (open eyes, aphakia, microphakia, hydronephrosis, and hydrocephalus). The maximum teratogenic response in mice was induced by treating pregnant mice intraperitoneally with 20 mg/kg cyclophosphamide on day 11 of gestation.

CYCLOPHOSPHAMIDE BIOTRANSFORMATION RATE AFFECTS ITS PRENATAL TOXICITY

The role of biotransformation in the production of cyclophosphamide teratogenicity was investigated in mice pretreated with either SKF-525A or phenobarbital (Gibson and Becker, 1968a). SKF-525A pretreatment of pregnant mice 30 minutes before cyclophosphamide significantly increased fetal weight loss and the incidence of resorptions, open eyes, polydactyly, aphakia, hydronephrosis, absence or nonossification of ribs, bone fusion, and bone curvature. Phenobarbital pretreatment of mothers for three days prior to cyclophosphamide, on the other hand, significantly reversed the cyclophosphamide-induced weight loss as well as incidence of limb defects, hydrocephalus, exencephaly, cleft-palate, aphakia, and hydronephrosis. In the maternal animal, plasma alkylating metabolites appeared more rapidly and reached higher levels as a result of phenobarbital pretreatment than in controls. In contrast, SKF-525A decreased the rate of appearance of alkylating metabolites in the maternal plasma. Phenobarbital pretreatment caused an overall increase in the activity of the drug metabolizing enzyme system and consequently increased cyclophosphamide activation. SKF-525A competed with cyclophosphamide for sites of metabolism in the enzyme system and decreased cyclophosphamide activation (Conney, 1967). Thus, the rate and extent of cyclophosphamide activation affected the incidence of congenital malformations.

These results suggested that cyclophosphamide teratogenicity was associated with the parent compound rather than alkylating metabolites (Gibson and Becker, 1968a; Gibson and Becker, 1971a,b). However, using C^{14}-labeled cyclophosphamide, pretreatment of pregnant animals with SKF-525A was found to increase the amount of cyclophosphamide reaching the embryo (Figure 2). The increased incidence of anomalies, therefore, was related to the increased amount of drug reaching the embryonic tissue (Gibson and Becker, 1971a). In contrast, phenobarbital pretreatment decreased in the amount of cyclophosphamide reaching the embryo and was correlated with a decrease in teratogenic effects.

Cyclophosphamide alkylating metabolites are more water soluble than the parent compound and the placenta is less permeable to these water soluble metabolites than to the relatively more lipid soluble parent compound (Gibson and Becker, 1971a). Thirty-two minutes after cyclophosphamide administration the proportion of parent compound present in the embryo was altered by pretreatment (Figure 2). Embryos from mothers pretreated with SKF-525A had a larger proportion of unchanged drug but embryos from mothers pretreated with phenobarbital had a larger proportion of drug metabolites. More drug reached the embryo in SKF-525A pretreated animals but phenobarbital pretreatment reduced the total amount of cyclophosphamide in the embryo. Because of the restricted placental transfer of polar metabolites the absolute level of metabolites present 32 minutes after cyclophosphamide was not affected by pretreatment (Figure 2). The pool of unmetabolized drug from which cyclophosphamide metabolites were generated, however, was much larger as a result of SKF-525A pretreatment (Figure 2) and apparently the embryo was capable of converting this unmetabolized drug to active alkylating metabolites. The incidence of cyclophosphamide anomalies, therefore, was increased.

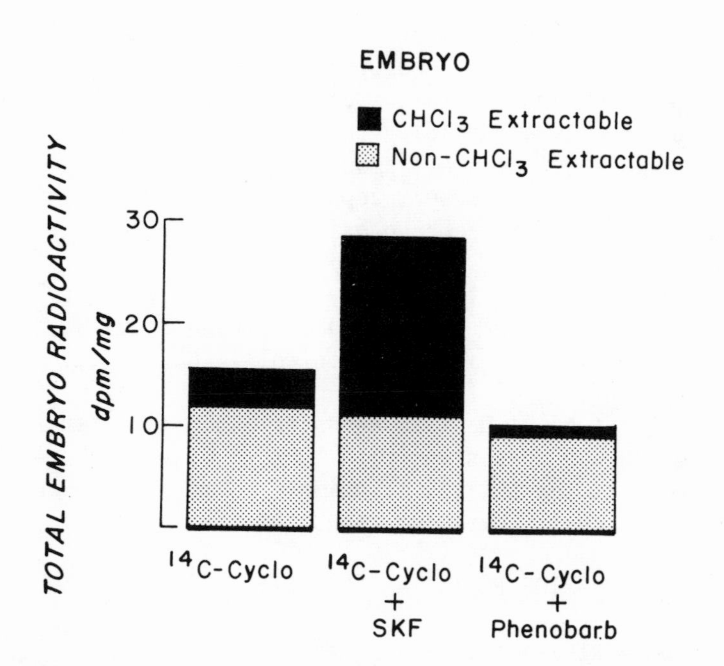

Figure 2. Chloroform extractable (parent compound) and non-chloroform extractable (metabolites) radioactivity from embryos of nonpretreated, SKF-525A-pretreated and phenobarbital-pretreated 11-day pregnant mice administered ^{14}C-cyclophosphamide. ^{14}C-cyclophosphamide, 20 mg/kg i.p., was administered and the embryos were removed 32 minutes later. SKF-525A, 32 mg/kg i.p., was administered one hour before cyclophosphamide and phenobarbital, 50 mg/kg i.p., was administered twice a day on gestational days 8, 9 and

10. Similar results were obtained with chloroform extraction of embryos 128 minutes after ^{14}C-cyclophosphamide.

IN VITRO METABOLISM OF CYCLOPHOSPHAMIDE IN EMBRYOS, FETUSES AND NEONATES

The activation of cyclophosphamide by the 9,000 x g supernatant of tissues from adult female mice and various perinatal tissues was studied and compared _in vitro_ under optimized conditions (Short and Gibson, 1971a). Relative to maternal liver neither the fetus nor the placenta was a major site for the generation of alkylating metabolites. The ability of the fetus and placenta to activate cyclophosphamide was only one-tenth the activity of maternal liver. In view of the teratogenicity and placental transfer data presented above, that small amount of cyclophosphamide activating ability, however, apparently was sufficient to cause cyclophosphamide embryotoxicity.

Cyclophosphamide activation by neonatal livers at 1, 3, 5, 7 and 14 days after birth ranged between only 10 and 30% of the activity found in adult controls. However, there was no significant difference in cyclophosphamide activation between livers of 21 day old neonates and mature females. This finding was consistent with that of other workers who had shown that the ability of animals to metabolize drugs increased to adult levels at about the time of weaning.

CYCLOPHOSPHAMIDE POSTNATAL TOXICITY

The administration of a single subcutaneous injection of cyclophosphamide to mice 24 to 48 hours after birth at a dose of 45 mg/kg significantly reduced the rate of growth and produced morphologic anomalies that were apparent at 49 days of age. Cyclophosphamide treated mice had delayed development of hair, short snouts, ears and tails, and reduced body weight at maturity. Cyclophosphamide neonatal toxicity was dose related but was not induced by equimolar doses of nor-nitrogen mustard, an agent with inherent alkylating activity (Short and Gibson, 1971b). In addition, mortality in the cyclophosphamide treated group over the period of 49 days was 30%. Thus, cyclophosphamide administration in a single dose at one day of age significantly decreased growth and produced morphologic anomalies. The dose of cyclophosphamide necessary to produce these effects was well below the dose of cyclophosphamide necessary to produce leukopenia, a sensitive measure of toxicity, in adults (Hayes *et al*., 1972).

The marked neonatal toxicity of cyclophosphamide was interesting since newborn animals had little ability to activate the drug to alkylating metabolites. Using ^{14}C-labeled cyclophosphamide, the disappearance of cyclophosphamide from plasma of the newborn proceeded by apparent first-order kinetics with a half-life of approximately 9 hours. In contrast, cyclophosphamide was eliminated from plasma of adults in a biphasic manner with a half-life during the initial phase of about 2 hours. These results were consistent with our understanding that metabolites of cyclophosphamide are more water soluble that the parent compound and therefore more readily excreted. The slow portion of disappearance curve represents the fraction of the alkylating agent which is bound to tissues. Thus, reduced metabolism of cyclophosphamide in the newborn results in less conversion of the parent compound to excretable metabolites in comparison to the adult. From these data we suggested that the increased toxicity of cyclophosphamide in the newborn was due to its decreased rate of elimination and continued release of toxic metabolites (Bus *et al*., 1973).

CYCLOPHOSPHAMIDE BIOTRANSFORMATION RATE AFFECTS ITS POSTNATAL TOXICITY

Phenobarbital was administered to pregnant mice twice a day during the last 3 days of gestation. This treatment had previously been shown to stimulate the "toxication-detoxication system" in the fetus (Hart *et al*., 1972). Animals delivered from pregnant females treated with phenobarbital late in gestation therefore had a stimulated "toxication-detoxication system". Phenobarbital stimulation of drug metabolism caused an increase in cyclophosphamide toxicity since the growth rate of animals treated with cyclophosphamide was depressed further than with cyclophosphamide alone (Figure 3). In addition, phenobarbital significantly increased the 49 day mortality of cyclophosphamide treated mice to about 70%. In other newborns the drug metabolizing enzyme system was inhibited by administering SKF-525A one hour before cyclophosphamide. SKF-525A inhibition of cyclophosphamide activation completely abolished the cyclophosphamide mortality and tended to reverse effects of cyclophosphamide on neonatal growth (Figure 3). SKF-525A and phenobarbital alone had no effect on the growth in newborn mice. Thus, stimulation of the cyclophosphamide activating system increased cyclophosphamide toxicity and inhibition of same system decreased cyclophosphamide toxicity.

Figure 3. Effect of phenobarbital (50 mg/kg i.p., twice daily to dam for three days immediately preceding delivery) and SKF-525A (32 mg/kg s.c., one hour prior to cyclophosphamide) pretreatment on the growth of newborn mice treated with cyclophosphamide (45 mg/kg s.c.). Each point is the mean mouse body weight within litters for five to six litters expressed as the percentage of adult control body weight.

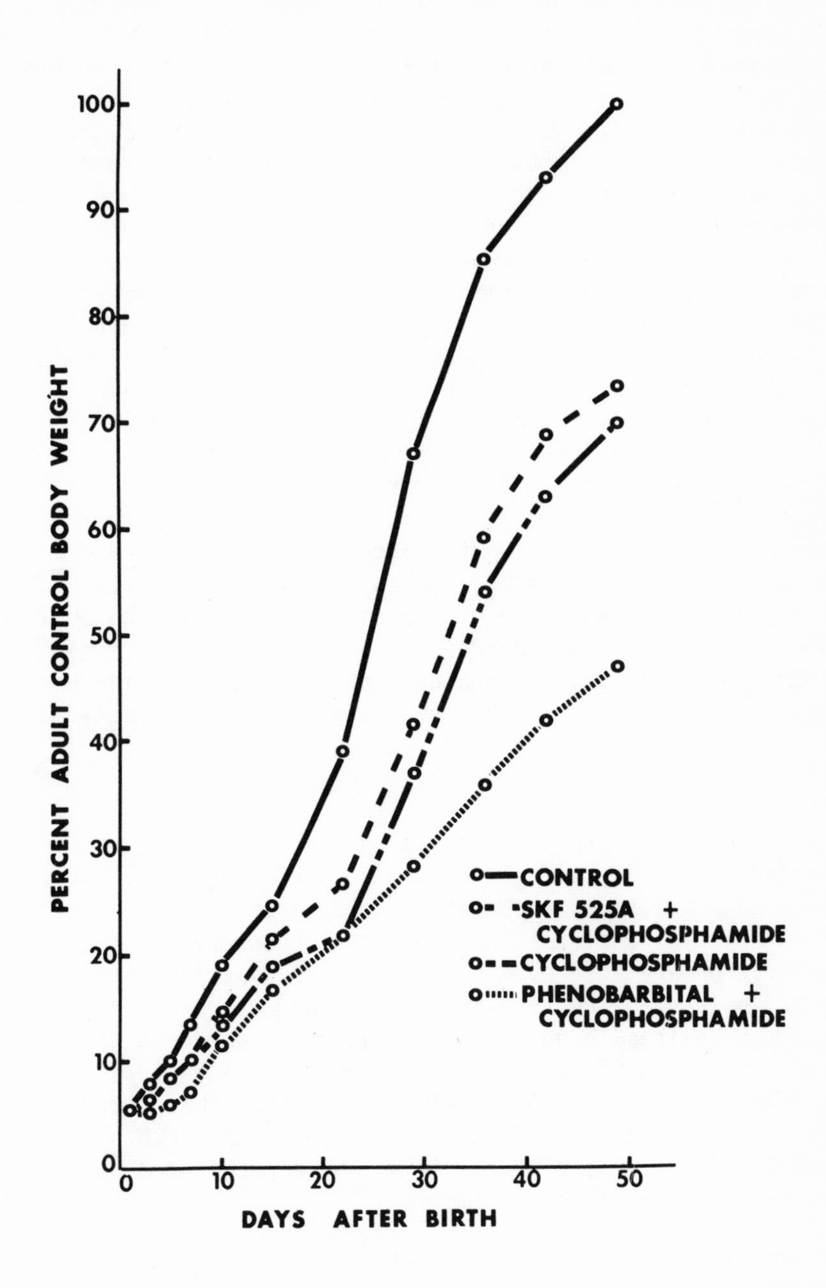
PERCENT ADULT CONTROL BODY WEIGHT
100
90
80
70
60
50
40
30
20
10
0
0
10
20
30
40
50
DAYS AFTER BIRTH
CONTROL
SKF 525A + CYCLOPHOSPHAMIDE
CYCLOPHOSPHAMIDE
PHENOBARBITAL + CYCLOPHOSPHAMIDE

Sixty-four minutes after cyclophosphamide administration to newborns 75% of the drug in plasma was the parent compound. In those animals receiving SKF-525A the percentage of unmetabolized drug was increased to 85% and in those animals pretreated with phenobarbital the amount of the parent compound was 50%. Thus, pretreatments increased or decreased the conversion of the parent drug to metabolites. The effect of these interventions on the rate of cyclophosphamide elimination, however, was not as dramatic as the change in toxicity. Pretreatment with SKF-525A increased the plasma half-life to nearly 12 hours but phenobarbital had no effect. Thus, in newborn animals the overall rate of elimination of cyclophosphamide and its metabolites was not as important as the rate of formation and levels of activated metabolites in the plasma. The phenobarbital-induced increase in toxicity correlated with the rapid formation and concurrently high plasma levels of alkylating metabolites whereas SKF-525A significantly decreased metabolite levels after cyclophosphamide administration. Thus, at a time when the cyclophosphamide activating ability in newborn mice was underdeveloped, stimulation of cyclophosphamide metabolism enhanced toxic effects and inhibition of cyclophosphamide metabolism prevented these effects.

CONCLUSION

These studies have demonstrated that the toxicity of cyclophosphamide, an agent which requires metabolic activation to alkylating metabolites, is subject to modification when the biotransformation processes necessary for that activation are altered. Changes in the biotransformation rate influenced the relative proportions of parent compound and metabolites in maternal plasma and altered the teratogenic response by influencing the availability of the chemical for placental transfer. Newborn animals were more vulnerable to toxic effects of cyclophosphamide than the adult and the extent of the toxicity was dependent on the extent of drug biotransformation.

These studies suggest that when the toxicity of any agent is determined in pregnant or newborn animals it may be prudent to consider that agent not only alone but also in combination with other drugs or chemicals. A large variety of drugs and environmental chemicals are capable of stimulating the "toxication-detoxication system" and thus, toxicity studies conducted in animals protected from these exposures may not appropriately represent the toxic response to be expected in man, domestic animals or wildlife. Similarly, other drugs and environmental chemicals produce inhibition of the "toxication-detoxication system" and these agents may also alter the action or toxicity of other drugs and chemicals. Unfortunately, it is not possible to test all combinations of drugs and environmental chemicals or industrial chemicals to determine their combined effects. However, it is important to understand that interactions can occur and we should be able to predict many of these on the basis of studies such as those described in this paper. Research into drug and chemical interactions in teratology and neonatal pharmacology-toxicology is in its infancy and these studies have demonstrated the role of altered biotransformation in perinatal chemical toxicity.

REFERENCES

Brock, N. and Hohorst, H.-J.: Ueber die Aktivierung von Cyclophosphamide *in vivo* und *in vitro*. Arzneimittel-Forschung 13: 1021-1031, 1963.

Brock, N. and Hohorst, H.-J.: Metabolism of cyclophosphamide. Cancer 20: 900-904, 1967.

Bus, J.S., Short, R.D. and Gibson, J.E.: Effect of phenobarbital and SKF-525A on the toxicity, elimination and metabolism of cyclophosphamide in newborn mice. J. Pharmacol. Exp. Ther. 184: 749-756, 1973.

Cohen, J.L. and Jao, J.J.: Enzymatic basis of cyclophosphamide activation by hepatic microsomes of the rat. J. Pharmacol. Exp. Ther. 174: 206-210, 1970.

Conney, A.H.: Pharmacological implications of microsomal enzyme induction. Pharmacol. Rev. 19: 317-366, 1967.

Conners, T.A., Grover, P.L. and McLoughlin, A.M.: Microsomal activation of cyclophosphamide *in vivo*. Biochem. Pharmacol. 19: 1533-1535, 1970.

Fields, R.B., Gang, M., Kline, I., Venditti, J.M. and Waravdekar, V.S.: The effect of phenobarbital or 2-diethylaminoethyl-2,2-diphenylvalerate on the activation of cyclophosphamide in vivo. J. Pharmacol. Exp. Ther. 180: 475-483, 1972.

Fouts, J.R.: Some studies and comments on hepatic and extrahepatic microsomal toxication-detoxication systems. Environmental Health Perspectives: experimental issue No. 2: 55-66, 1972.

Fouts, J.R. and Adamson, R.H.: Drug metabolism in the newborn rabbit. Science (Washington) 129: 857-858, 1959.

Fouts, J.R. and Devereux, T.R.: Developmental aspects of hepatic and extrahepatic drug-metabolizing enzyme systems: Microsomal enzymes and components in rabbit liver and lung during the first month of life. J. Pharmacol. Exp. Ther. 183: 458-468, 1972.

Gerlinger, P.: Action du cyclophosphamide injecté à la mére sur la réalisation de la forme du corps des embryons de lapin. C. R. Seances Soc. Biol. Filiales 158: 2154-2157, 1964.

Gibson, J.E. and Becker, B.A.: Effect of phenobarbital and SKF-525A on the teratogenicity of cyclophosphamide in mice. Teratology 1: 393-398, 1968a.

Gibson, J.E. and Becker, B.A.: The teratogenicity of cyclophosphamide in mice. Cancer Res. 28: 475-480, 1968b.

Gibson, J.E. and Becker, B.A.: Effect of phenobarbital and SKF-525A on placental transfer of cyclophosphamide in mice. J. Pharmacol. Exp. Ther. 177: 256-262, 1971a.

Gibson, J.E. and Becker, B.A.: Teratology of structural truncates of cyclophosphamide in mice. Teratology 4: 141-151, 1971b.

Goldenthal, E.I.: A compilation of LD50 values in newborn and adult animals. Toxicol. Appl. Pharmacol. 18: 185207, 1971.

Hackenberger, I., Kreybig, T. von: Vergleichende teratologische Untersuchungen bei der Maus und der Ratte. Arzneimittel-Forschung 15: 1456-1460, 1965.

Hart, L.G., Adamson, R.H., Dixon, R.L. and Fouts, J.R.: Stimulation of hepatic microsomal drug metabolism in the newborn and fetal rabbit. J. Pharmacol. Exp. Ther. 137: 103-106, 1962.

Hayes, F.D., Short, R.D. and Gibson, J.E.: A correlation between cyclophosphamide induced leukopenia in mice and the presence of alkylating metabolites. Proc. Soc. Exp. Biol. Med. 139: 417-421, 1972.

Henderson, P.T.: Metabolism of drugs in rat liver during the perinatal period. Biochem. Pharmacol. 20: 1225-1232, 1971.

Hill, D.L., Laster, W.R., Jr. and Struck, R.F.: Enzymatic metabolism of cyclophosphamide and nicotine and production of a toxic cyclophosphamide metabolite. Cancer Res. 32: 658-665, 1972.

Jondorf, W.R., Maickel, R.P. and Brodie, B.B.: Inability of newborn mice and guinea pigs to metabolize drugs. Biochem. Pharmacol. 1: 352-354, 1959.

Kreybig, T. von: Die teratogene Wirkung von Cyclophosphamide wahrend der embryonalen Entwicklungsphase bei der Ratte. Naunyn-Schmeidegergs Arch. Pharmakol. 252: 173-195, 1965.

MacLeod, S.M., Renton, K.W. and Eade, N.R.: Development of hepatic microsomal drug-oxidizing enzymes in immature male and female rats. J. Pharmacol. Exp. Ther. 183: 489-498, 1972.

Murphy, M.L.: Teratogenic effects of growth inhibiting chemicals, including studies on thalidomide. Clin. Proc. Child. Hosp. D.C. 18: 307-322, 1962.

Rauen, H.M. and Kramer, K.P.: Der Gesamtalkylantien-Blutspiegel von Ratten nach Verabreichung von Cyclophosphamid and Acyclophosphamid. Arzneimittel-Forschung 14: 1066-1067, 1964.

Roberts, J.J., Brent, T.P. and Crathorn, A.R.: The mechanism of the cytotoxic action of alkylating agents on mammalian cells. In P. N. Campbell (ed.), The Interaction of Drugs and Subcellular Components in Animal Cells, pp. 5-27, London: J. & A. Churchill, Ltd., 1968.

Shoji, R. and Ohzu, E.: Effect of endoxan on developing mouse embryos. J. Fac. Sci. Hokkaido Univ. Serv. VI Zool. 15: 662-665, 1965.

Short, C.R. and Davis, L.E.: Perinatal development of drug-metabolizing enzyme activity in swine. J. Pharmacol. Exp. Ther. 174: 185-196, 1970.

Short, R.D. and Gibson, J.E.: Development of cyclophosphamide activation and its implications in perinatal toxicity to mice. Toxicol. Appl. Pharmacol. 19: 103-110, 1971a.

Short, R.D. and Gibson, J.E.: The effects of cyclophosphamide and nor-nitrogen mustard administration to one-day-old mice. Experientia (Basel) 27: 805-806, 1971b.

Short, R.D., Rao, K.S. and Gibson, J.E.: The in vivo biosynthesis of DNA, RNA, and proteins by mouse embryos after a teratogenic dose of cyclophosphamide. Teratology 6: 129-137, 1972.

Short, R.D. and Gibson, J.E.: Biosynthesis of deoxyribonucleic acid, ribonucleic acid and protein in vivo by neonatal mice after a toxic dose of cyclophosphamide. Biochem. Pharmacol. 22: 3181-3188, 1973.

Sladek, N.E.: Therapeutic efficacy of cyclophosphamide as a function of its metabolism. Canc. Res. 32: 535-542, 1972.

Soyka, L.F.: Determinants of hepatic aminopyrine demethylase activity. Biochem. Pharmacol. 18: 1029-1038, 1969.

Wheeler, G.P. and Alexander, J.A.: Effects of nitrogen mustard and cyclophosphamide upon the synthesis of DNA in vivo and in cell free preparations. Canc. Res. 29: 98-109, 1969.

Wilson, J.G.: Teratogenic interaction of chemical agents in the rat. J. Pharmacol. Exp. Ther. 144: 420-436, 1964.

Wilson, J.G.: Embryological considerations in teratology. In J. G. Wilson and J. Warkany (eds.), Teratology Principles and Techniques, pp. 251-278, Chicago: The University of Chicago Press, 1965.

Wilson, J.T.: Alteration of normal development of drug metabolism by injection of growth hormone. Nature (London) 225: 861-863, 1970.

THE USE OF ELECTROCHEMICAL TECHNIQUES IN FDA ANALYTICAL PROCEDURES

Albert L. Woodson

Food and Drug Administration
Chicago District Laboratory, Room 1222
Post Office Building, Chicago, Ill. 60607

Today's analytical procedures utilize a variety of techniques to identify and quantitate the ion, radical or compound under investigation. These techniques may be divided into two general classes: (1) Optical, and (2) Electrochemical.

In the past, the optical technique was usually the technique of choice. This was primarily due to the availability of commercial instrumentation using optical detection techniques. Commercial instruments using electrochemical methods of detection have not been as profuse as the optical instruments. However, this does not mean that optical methods are superior to electrochemical methods. This may be true in some instances, while the reverse may be the case in other instances. The abundance of optically oriented commercial instruments merely indicates that manufacturers considered optically oriented instrumentation to be the most profitable.

A survey of analytical instruments, currently available, indicates a trend toward an increasing application of electrochemical techniques to the identification and quantitation of ions, radicals or compounds.

The use of an electrochemical method in any analytical technique requires that the ion, radical or compound under investigation undergoes a change due to gain or loss of one or more electrons. The device in which this transformation takes place is the electrochemical cell. This device consists of at least two conducting electrodes placed in a conducting solution. The electrochemical cell possesses a variety of concentration dependent physical characteristics, which may be exploited for chemical analysis. Consequently, the number of experimental methods based on the electrochemical cell are many.

There are, basically, two physical characteristics of the electrochemical cell which control the function of the cell: (1) The cell equilibrium potential--a thermodynamic property, and (2) the cell impedance--a kinetic property. The cell impedance is a combination of the bulk impedance, represented by the total ohmic resistance of the cell solution, and an interfacial impedance. The interfacial impedance consists of two parts: (1) The double-layer impedance, and (2) the faradaic impedance. The double-layer acts like a capacitor, permitting current flow by the charge transfer associated with charging and discharging of the double-layer. The faradaic impedance accommodates current flow by the actual charge transfer associated with the electrochemical reaction taking place.

Analytical methods utilizing an electrochemical cell may be classified according to the physical characteristics of the cell that gives rise to useful experimental data. For example, a potentiometric titration is responsive to the cell equilibrium potential. The various types of polarography and linear scan voltammetry are examples of techniques that are responsive to cell impedance.

Normally, two or three electrodes are used in electrochemical techniques. One electrode is a reference electrode, whose potential is known and constant; the second electrode is the working or indicator electrode; the third electrode, if used is an auxiliary electrode, usually a piece of platinium wire. The working electrode potential depends upon the type of technique being used; it may or may not be varied. The auxiliary electrode permits the use of high resistance organic solvents as non-aqueous media. Reference electrodes usually used are the SCE and Ag/AgCl electrodes. The working electrode may be a solid or dropping mercury electrode. Solid electrodes are usually constructed from Pt, Au, or graphite. The hanging mercury drop is considered a solid electrode.

Solid electrode techniques are: (1) Linear Scan Voltammetry, and (2) Anodic Stripping Voltammetry. Dropping mercury electrode techniques are the various types of polarography.

As a regulatory agency, the Food and Drug Administration must investigate and apply any new analytical technique that will enable it to fulfill its responsibility to the public. In the past, FDA analytical procedures using instrumentation in the determinative step have, mainly, used spectrophotometric and chromatographic instruments. Currently, electrochemical techniques are being used as the determinative step in the development of analytical procedures. These techniques are: (1) Linear Scan Voltammetry, (2) Various polarographic techniques, and (3) Anodic Stripping Voltammetry.

When a linearly varying potential is applied to a stationary electrode, such as the hanging mercury drop, a current-voltage relationship is obtained. If the potential of the electrode is rapidly scanned linearly, any electroactive species having a redox potential in the range scanned will cause a flow of current. This produces a peak-shaped curve. The technique is called Linear Scan Voltammetry and has a sensitivity of about 10^{-6}M. The analytical aspects of this curve are: (1) the potential at which the peak current occurs, and (2) the magnitude of the peak current. The potential producing the peak current is a qualitative parameter, while the magnitude of the current is a quantitative parameter.

If we substitute a dropping mercury electrode for the hanging drop of solid electrode, used in Linear Scan Voltammetry, we have a special type of voltammetry. This special type is given the name polarography.

Several advantages peculiar to the dropping mercury electrode are responsible for the more widely acceptance of polarography over other voltammetric techniques. The most important advantage is that, except under certain very unusal circumstances, each drop exactly duplicates the behavior of its predecessor. Consequently, the currents are accurately reproducible from one drop to the next and independent of the previous history of the experiment. Solid products cannot accumulate on the electrode surface, changing its properties, as is possible with solid electrodes. The duration of the experiment is practically without effect, leaving only three important variables (electrode potential, solution composition, and current) to be considered.

There are two main disadvantages associated with the dropping mercury electrode. One is the limited useful potential range. Since mercury is easily oxidized, very positive oxidizing potentials cannot be secured. The other disadvantage is that the continuous variation of the electrode area gives rise to significant currents even in the absence of a reducible or oxidizable substance.

In FDA laboratories instrumentation is available for performing polarographic analysis in six different operational modes. These are: (1) Conventional DC, (2) Sampled DC or Tast, (3) Pulse, (4) Differential Pulse, (5) Dual Cell or Differential, and (6) Alternating Current Polarography.

The polarogram obtained by conventional DC polarography consists of two current components, the desired faradaic current and the undesired double-layer charging current. This double-layer charging current limits the analytical sensitivity. It arises from drop growth and appears on the polarogram as the individual drop oscillations. Any increase in the analytical sensitivity of conventional DC can only be accomplished by providing some means of measuring the faradaic response, while discriminating against the double-layer response. Such an arrangement would enable one to measure small faradaic responses that would otherwise be lost in the magnified double-layer current, at extreme sensitivity settings of the recorder.

Electronic circuits have been devised that will minimize the contribution of the double-layer charging current and the resulting polarogram will exhibit the response of the DME to increases in the faradaic current, which is a measure of the concentration of the electroactive species. An electronic sample-and-hold circuit and a controlled drop timing circuit accomplishes this desired effect. The polarogram obtained is identical in shape to the conventional DC polarogram, but minus the recorded individual drop oscillations. This technique known as Sampled DC or Tast polarography extends the conventional DC limit of detection from 10^{-5}M to about 10^{-6}M.

Pulse polarographic techniques offer tremendous advantages over conventional DC in terms of sensitivity and resolution. There are essentially two pulse polarographic techniques; normal pulse and differential pulse. In normal pulse polarography, potential pulses of successively increasing amplitude are applied from a fixed initial potential to the working electrode at a fixed time during the drop life. The pulses are usually 50-60 mS in duration and the current is sampled at some fixed time (usually 40 mS) after pulse application. The time between pulses is usually 0.5 - 5S, and sample and hold circuits are utilized to display the sampled currents on a recorder. The resulting polargram has the same shape as the Sampled DC wave, but the limit of detection is lowered to about 10^{-7}M.

In differential pulse polarography, a slowly varying potential ramp is applied to the working electrode exactly as in DC polarography. However, superimposed on this ramp are constant amplitude pulses of short duration synchronized to occur at the end of the mercury drop life. The current is sampled twice during the life of each drop; once just before the pulse application and again at a predetermined time after pulse application (about 40 mS). The difference between these measurements is obtained electronically and displayed on a recorder. A peak-shaped curve is obtained and the limit of detection is lowered to about 10^{-10}M.

Alternating current polarography is performed by controlling all experimental variables in a manner identical to DC polarography while superimposing on the linear potential sweep a small amplitude sinusoidual potential of fixed frequency and amplitude. The AC polarogram is a recording of the resulting alternating currents as a function of the DC potential. This produces a peak-shaped curve. The sinusoidal signals of AC polarography can be subjected to electronic tricks in order to minimize the double-layer charging contribution. Applying these tricks, such as phase-sensitive AC detection and the 2nd harmonic of the fundamental frequency of the sinusoidal signal, the detection limit may be lowered to about 10^{-10}M.

Dual cell or differential polarography makes use of two cells each containing a DME maintained at the same potential. If different solutions are placed in each cell, the difference between the currents flowing through them is recorded as a function of their common potential. This technique requires that the two DME are as nearly identical as possible. Any slight variation in capillary characteristics may be compensated for by the electrode balance adjustment of the instrument. The above mentioned operation modes of polarography may be carried out in either a dual or single cell operation.

The detection of metals in trace amounts can be easily accomplished through the use of the previously discussed polarographic techniques and controlled potential electrolysis. If a solution containing trace amounts of a metal ion is electrolyzed at a potential on the plateau of the polarographic wave of the metal, the reduction product will accumulate at the electrode surface. If the electrode is a hanging mercury drop electrode, the metal, formed by the reduction of the metal ion, forms an amalgam with the mercury drop. Prolonging the period of electrolysis causes a concentration of the metal within the drop. Oxidation of the deposited metal yields a peak current proportional to the amount of the metal concentrated in the mercury drop. This results in considerable improvement in sensitivity and lower detection limits. This technique is called Anodic Stripping Voltammetry.

Since the formation of a mercury amalgam is a prerequisite, to the accumulation of the reduction product on the electrode surface, only metals forming amalgams with mercury may be determined using this technique with the DME. However, if a Pt electrode is used, ions such as As, Se, Au, Ag, and even Hg may be detected by this technique. As and Se do not readily form mercury amalgams and the oxidation potentials of Au and Ag are too positive to effectively use the hanging mercury drop as the working electrode. A Pt electrode overcomes these problems, since As, Se, Au, Ag and Hg are easily deposited on a clean Pt surface. Also, the use of Pt extends the positive potential range, allowing the easy re-oxidation of the elements deposited during the electrolysis step. The main disadvantage of using Pt as the working electrode is the necessity of providing a clean electrode surface for successive determinations. With the hanging mercury drop, a clean surface is easily obtained by hanging a fresh drop. The hanging mercury drop electrodes, currently available, have a micrometer attachment which provides a clean fresh drop of identical size by merely turning the micrometer a predetermined number of scale divisions.

The so-called WIG electrode (a wax-impregnated graphite electrode) can be used instead of the Pt electrode. It is easier to provide a fresh, clean surface with this electrode, than with the Pt electrode. However, in some cases the WIG electrode reduces the sensitivity of detection.

The anodic scan of the sweep potential may be applied in either the DC, pulse or AC mode. The degree of sensitivity varies with the mode of operation, the DC mode being the least sensitive. Detection limits have been obtained in the ppb range with differential pulse of AC modes of operation.

Regardless of the mode of the anodic scan, the recording is a peak-shaped curve. The magnitude of the peak current is controlled by the following parameters: (1) the size of the hanging drop, (2) the rate of stirring during the electrolysis, (3) the time period for electrolysis, and (4) the rate of the anodic scan. Increasing one or more of these parameters increases the magnitude of the peak current. However, if these parameters are kept constant, the magnitude of the peak current is proportional to the trace element concentration.

In FDA laboratories, anodic stripping voltammetry is being investigated because its limits of detection for most trace metals are below those obtainable with the conventional AA technique. By conventional AA technique, I mean one that does not employ procedures that serve to concentrate the desired trace metal before it is atomized or sprayed into the flame. Examples of such techniques are (1) the use of organic chelating agents, (2) the Delves Cup technique for Pb determination, and (3) the carbon red technique.

Another advantage of anodic stripping voltammetry over AA is the ability to perform trace analysis of one or more metals, simultaneously, in the same solution. To perform this analysis by AA, separate aliquots of the solution must be used for each metal. With anodic stripping voltammetry, the potential of the electrolysis may be selected to include certain metals. Consider a mixture of Cu, Pb, Cd, and Zn. The half-wave potentials of these ions in 1 N HCl increases negatively from Cu to Zn. The half-wave potential of Zn is -1.0 volt. If this mixture is electrolyzed at a potential of -0.8 volt, only Cu, Pb and Cd will be reduced and deposited in the mercury drop. On anodic scanning, they will be stripped from the drop in the order of decreasing negative potential. That is, the metal having the most negative half-wave potential will be stripped first. The recorded scan will show a peak for Cd, Pb, and Cu, in that order.

In FDA laboratories, anodic stripping voltammetry is being used in the determination of trace elements in foods; specifically, Pb, Cd, and Zn in fish, and Pb in evaporated milk.

The overall appeal of polarography as an analytical technique is enhanced by its application to the analysis of organic compounds. This situation arises because an extremely large number of organic functional groups are either directly reducible or oxidizable at the working electrode, or can be readily and cleanly functionalized to yield electroactive compounds.

The official compendia of FDA, the USP and the NF, contain seven drug monographs in which the assay procedure is polarographic (1).

The insolubility of some drugs in aqueous media does not preclude their polarographic capabilities, since polarography may be performed in aqueous media. The use of the three electrode cell and current polarographic instrumentation makes possible the use of high resistance organic solvents as polarographic supporting electrolytes.

The DC and AC polarographic response of 24 drugs in nonaqueous media has been investigated in our laboratory (2). The compound classifications included in this work are: Barbiturates, Salicylates, Corticosteroids, Alkaloids, Sulfa Drugs, Nitro compounds and Estrogens. The majority of the compounds studied yielded ideal responses for analytical purposes.

Figures 1 and 2 show the response of reserpine in nonaqueous media. The concentration of reserpine is 1.73×10^{-4}M in 0.1M tetrabutylammonium perchlorate in acetonitrile.

Non-aqueous organic polarography was used in our laboratory to demonstrate the sensitivity of the polarographic response of nitroglycerin (3). A single component formulation containing 0.6 mg/tab of nitroglycerin was assayed by non-aqueous polarography, the A.O.A.C. infrared and phenoldisulfonic acid methods (4), and the method of Hohmann and Levine (5). The polarographic assay was performed without prior isolation of the nitroglycerin, by, simply adding the supporting electrolyte to the powdered tablet material. The supporting electrolyte was the organic solvent, acetonitrile, containing 0.1M tetrabutylammonium perchlorate. All of the inorganic components of the formulation were insoluble; therefore, after 15 minutes the mixture was filtered and the filtrate added to the polarographic cell. Figure 3 shows the DC polarographic response of the extracted nitroglycerin.

The mean value of five determinations of a composite sample was 0.607 mgs/tab, or 101.2% of the tablet declaration.

The IR, phenoldisulfonic acid, and the Hohmann and Levine methods required the isolation of the nitroglycerin before applying the determinative step. The values obtained by these procedures were: IR--0.639 mg/tab, or 106.5%; phenoldisulphonic acid--0.651 mg/tab, or 108.5%; and Hohmann and Levine--0.620 mg/tab, 103.3%.

Five individual tablets of the single component formulation were assayed polarographically. The mean value was 0.612 mg/tab, or 102%. This agrees well with the mean value of 0.607 mg/tab obtained with the composite polarographic assay.

A multiple component formulation containing per tablet: phenobartital--16 mg, nitroglycerin--0.26 mg, sodium nitrite--65 mg, and veratum viride--48.5 mg was assayed (1) polarographically, (2) by the method of Hohmann and Levine, and (3) by the A.O.A.C. IR method. The presence of sodium nitrite in the formulation preclued the use of A.O.A.C. phenoldisulphonic acid method.

Six polarographic determinations of a composite sample gave a mean value of 0.035 mg./tab (13.4%); 5 individual tablet polarographic determinations gave a mean value of 0.035 mg./tab. No results were obtained with the IR procedure, because of the low nitroglycerin content. The method of Hohmann and Levine, using the composite sample, gave a value of 0.042 mg./tab (16.1%).

The low assay values for nitroglycerin suggest decomposition of nitroglycerin within the tablet. To insure

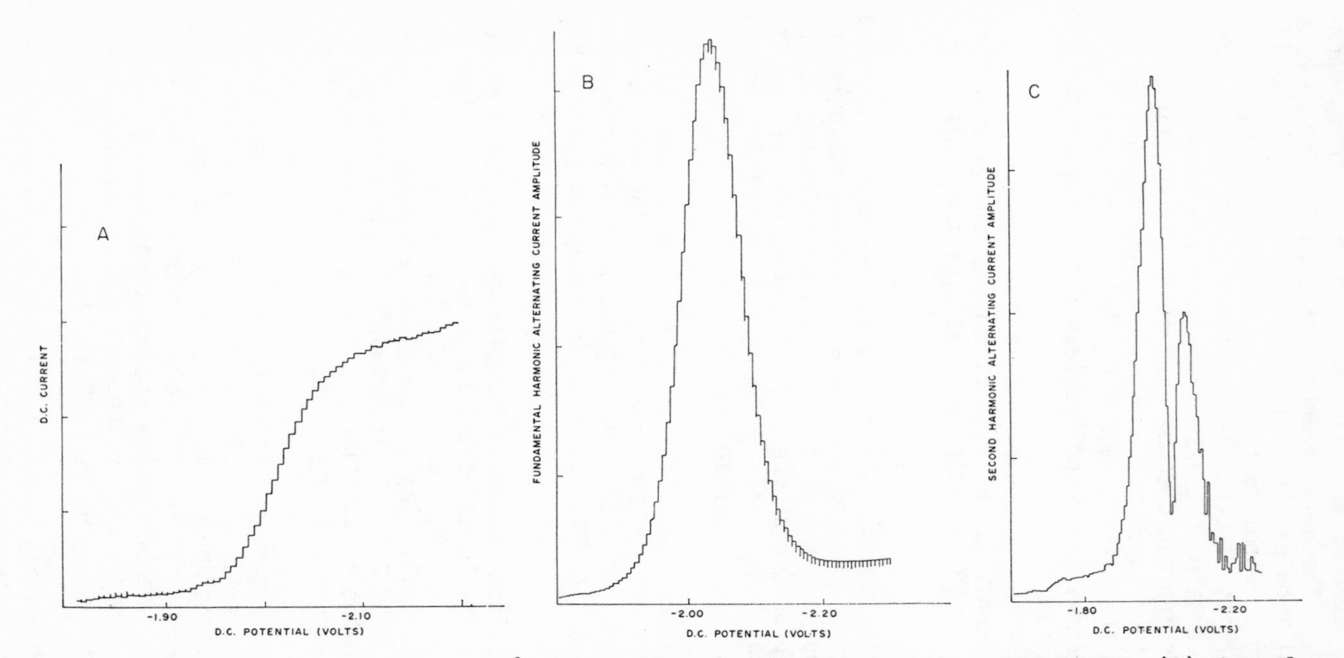

Figure 1. Polarograms of 2.79 x 10^{-4}M testosterone in acetonitrile-0.10M TBAP. (A) dc polarogram; (B) fundamental harmonic ac polarogram; (C) second harmonic ac polarogram (ordinates uncalibrated in all cases).

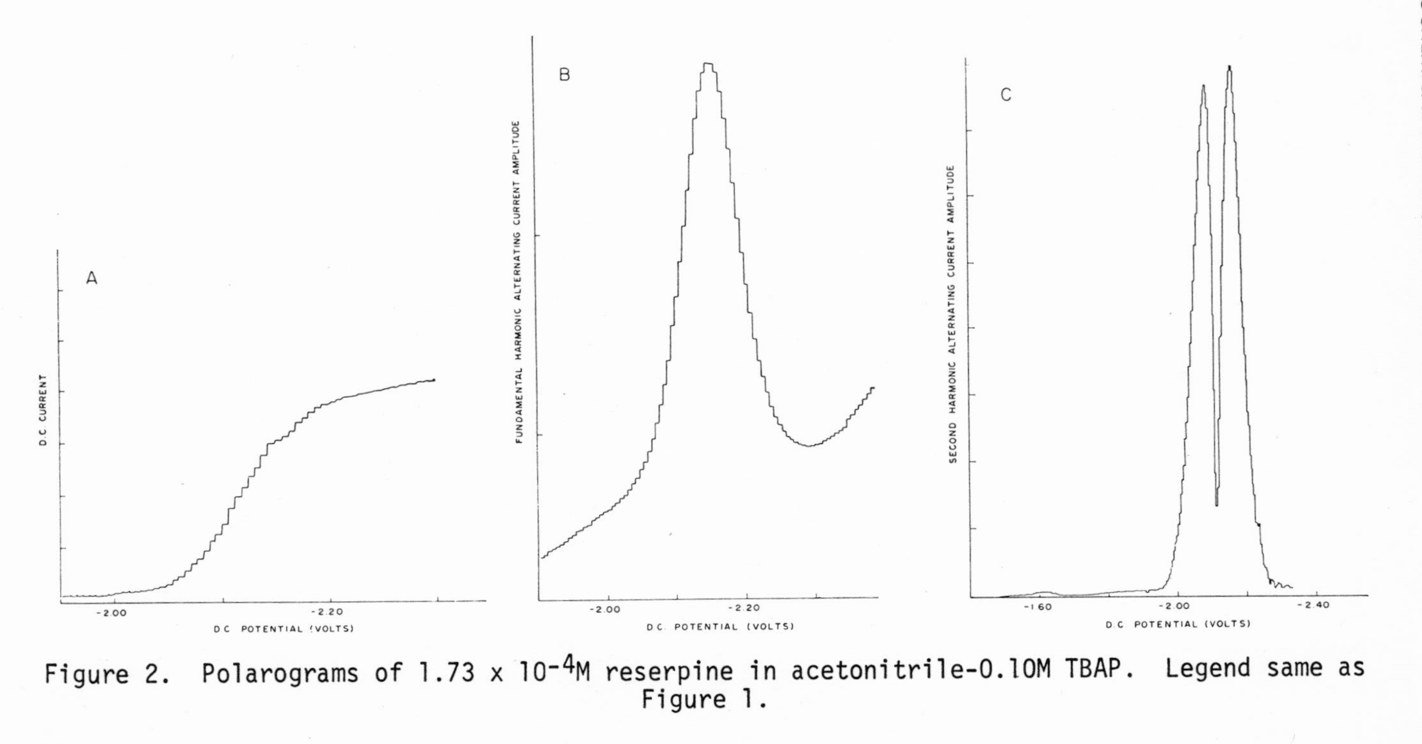

Figure 2. Polarograms of 1.73×10^{-4}M reserpine in acetonitrile-0.10M TBAP. Legend same as Figure 1.

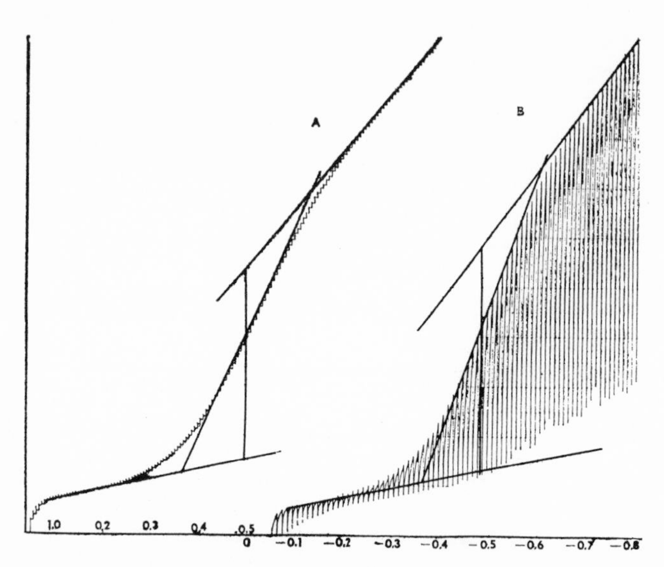

Figure 3. DC polarogram of nitroglycerin (0.060 mg/ml); supporting electrolyte 0.1M tetrabutylammonium perchlorate in acetonitrile. (A) recorded with sample-and-hold circuit; (B) recorded without sample-and-hold circuit.

that low values were not due to the technique, recovery assays were made. Known amounts of standard nitroglycerin were added to weighed portions of the assayed composite sample and the mixture polarographically analyzed. Five determinations gave a percentage recovery range of 89.1-102.7%, with a mean value of 95.5%.

Figure 4 shows the DC polarographic response of this sample. The effect of phenobarbital and veratum viride on the half-wave potential of nitroglycerin and the slope of the nitro limiting current is shown. Veratum viride, alone, did not produce a polarographic wave; however, the half-wave potential of nitroglycerin shifts to a more negative potential in the presence of this alkaloid.

To demonstrate the linearity of wave-heights versus nitroglycerin concentration, in the presence of phenobarbital and veratum viride, a standard curve was prepared by assaying solutions of nitroglycerin, phenobartital and viride, when only the nitroglycerin concentration was varied.

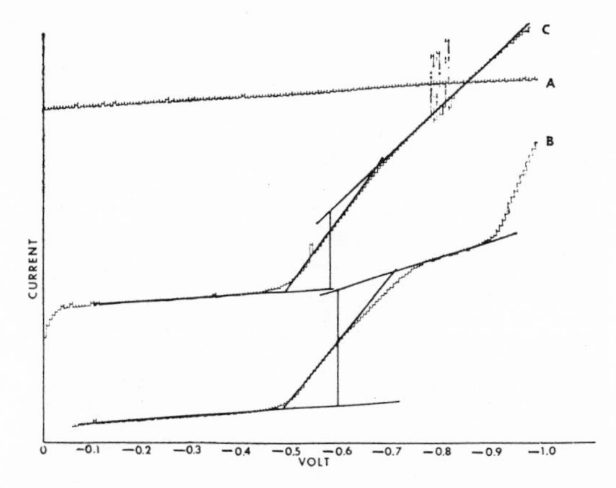

Figure 4. Effect of veratum viride and phenobarbital on half-wave potential and the slope of the limiting current of the nitroglycerin wave.

Figure 5 shows the standard curve obtained; also, the standard curve obtained using only nitroglycerin standard solutions. Note the slope of Curve A. This is due to the presence of phenobarbital in the solution being polarographed.

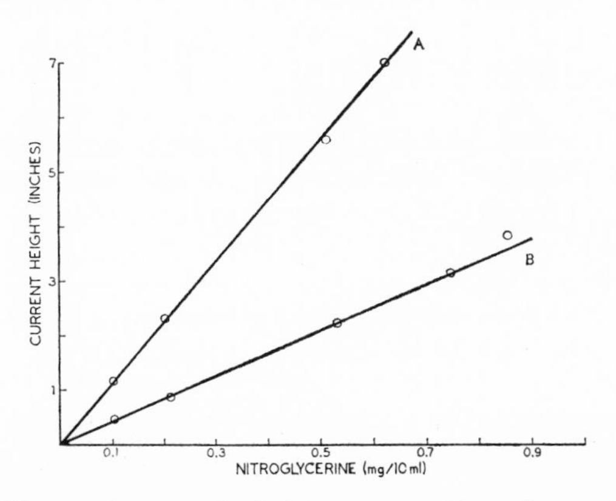

Figure 5. Plot of wave heights vs. nitroglycerin concentration. (A) mixture of nitroglycerin, veratum viride, and phenobarbital; (B) nitroglycerin.

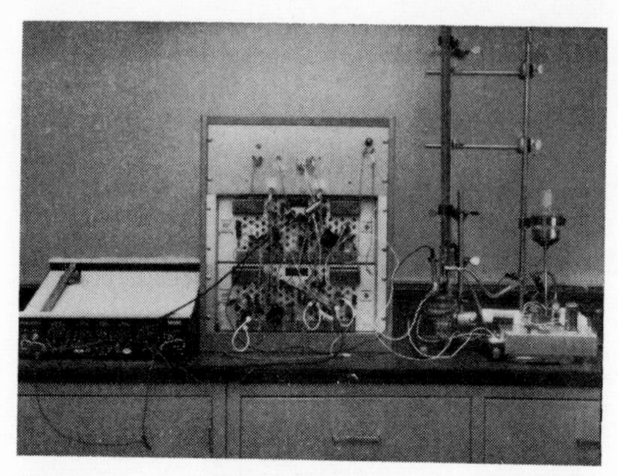

Figure 6. Polarographic instrument.

Figure 6 shows the polarographic instrument used in this work. It was constructed in the Chicago-District FDA laboratories.

REFERENCES

1) The United State Pharmacopeia, Eighteenth Revision. The National Formulatory, Thirteenth Edition.

2) Albert L. Woodson and Donald E. Smith, Direct Current and Alternating Current Polarographic Response of Some Pharmaceuticals in an Aprotic Organic Solvent System, Anal. Chem. 42, 242-48 (1970).

3) Albert L. Woodson and Larry L. Alber, Non-Aqueous Polarographic Analysis of Nitroglycerin, J.A.O.A.C., 52, 847-53 (1969).

4) Official Methods of Analysis, 10th Ed., Association of Official Agricultural Chemists, secs. 32.320-32.325.

5) J. R. Hohmann and J. Levine, Partition Column for Determination of Glyceryl Trinitrate in Tablets, J.A.O.A.C., 47, 471-73 (1964).

6) Louis Meites, Polarographic Techniques, Second Edition, Interscience Publishers.

ATOMIC ABSORPTION IN FOOD ANALYSIS —
SPECIAL TECHNIQUES FOR TRACES OF HEAVY METALS

Walter Holak

FDA-DHEW
850 Third Avenue
Brooklyn, New York 11232

CONVENTIONAL ATOMIC ABSORPTION SPECTROPHOTOMETRY

Atomic absorption spectrophotometry (AAS) provides one of the most useful and convenient means for the determination of metallic elements in solution from a wide variety of sample types. The basic requirement is that the element of interest be solubilized either in aqueous or certain organic solvents for subsequent aspiration into the flame, whereby the atomic absorption (AA) signal is measured. This general sampling technique is perfectly adequate for most applications down to the parts per million level. At the sub parts per million level some form of sample preconcentration and clean-up may be necessary. This is usually accomplished by chelation and extraction into an organic solvent.

In the analysis of foods for metallic elements, if the sample is a beverage it can often be aspired directly and the atomic absorption signal measured. Other food sample types such as animal tissues, plants and vegetables must first undergo a sample pre-treatment step. This means that the sample must be ashed at a temperature no higher than 500°C to prevent losses of some of the more volatile elements. The sample can also be digested using mixtures of nitric, sulfuric and perchloric acids.

At the present time we are particularly interested in traces of heavy metals in foods because of their widespread occurrence in the environment, as a result of industrial pollution, and their toxic effects at low concentrations on biological systems. Some of the metals of concern are shown in Table 1.

For many of the metals listed, at very low concentrations (0.01 ppm) we have found conventional atomic absorption spectrophotometry inadequate for their determination in foods because of the lack of sensitivity or interferences. For example, the detection limit for mercury in aqueous solution is about 0.01 ug/ml; we wish to quantitate mercury even below this level in some food samples. For this reason we resort to the much more sensitive Hatch and Ott cold vapor technique (1). Elements whose resonance lines lie in the far ultra-violet region of the spectrum suffer severely from non-specific background absorption interference. This includes lead, cadmium and zinc, among others. The difficulty in the arsenic and selenium determinations is even more serious because the resonance lines of these elements lie below 200 nm where flame components also absorb strongly.

Table 1. Several Heavy Metals Of Interest

	Residues Found (Results In PPM)		
Determinative Step ⇨	Atomic Absorption	Photometric	GLC
Mercury			
Lead			
Arsenic			
Cadmium			
Selenium			
Chromium			
Cobalt			
Nickel			
Zinc			
Antimony			
Copper			
Tin			
Manganese			

I would like to cite one practical example of the difficulty encountered in the determination of lead in evaporated milk by conventional atomic absorption spectrophotometry where background absorption was overlooked. In recent months there have been reports in the press of findings by some laboratories of significant levels (1 ug/g) of lead in evaporated and condensed milk. The method consisted of ashing a relatively large sample (100 g) and dissolving the residue in a small volume of dilute hydrochloric acid. The resultant solution was then aspirated into a sir-acetylene flame of an atomic absorption spectrophotometer. The signal was measured without any regard to the background absorption. The effect is shown in Figure 1. The absorption of the lead standard (1 ug/ml) and of the sample solution are measured at the lead resonance line of 283.7 nm using a fairly large scale expansion. The absorption of the sample solution was roughly one-half of the standard corresponding to approximately 0.5 ug/ml lead. Absorptions of the same solutions under the same conditions were also measured at 280 nm, several nanometers away from the lead resonance line. There is no absorption due to the lead standard but the sample still absorbs as much as before. This indicates that some other species in the sample other than lead are giving rise to almost the entire signal.

The preceding example illustrates a very common problem encountered in the analysis of foods for traces of heavy metals using atomic absorption spectrophotometry. As the size of the sample is increased and the volume in which the sample is dissolved is kept small, so as to increase the detection limit, the background absorption becomes significant and must be taken into account or the analytical result will be in serious error. For this and other reasons special techniques have been developed to circumvent the difficulties encountered in conventional atomic absorption spectrophotometry. This is the main thrust of this writing. The subject areas which I will consider are: (1) the destruction of organic matter and the dissolution of sample; (2) the gas-sampling techniques using metal hydrides and metal chelates; (3) the use of the Sampling Boat in food analysis as utilized in FDA's New York District.

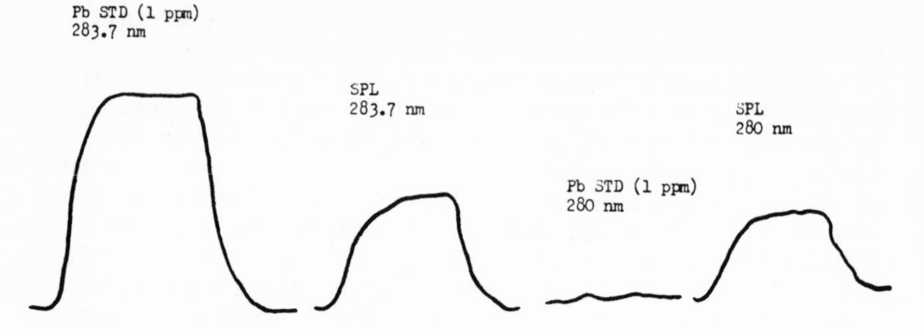

Figure 1. Example of the background interference in the determination of lead in evaporated milk by conventional AAS.

DESTRUCTION OF ORGANIC MATTER AND DISSOLUTION OF SAMPLE - THE TEFLON VESSEL

There are numerous methods for the decomposition of organic samples for subsequent elemental analysis by atomic absorption spectrophotometry. Most procedures utilize mixtures of strong acids with oxidizing properties in open vessels. However, there are essentially two problems associated with the use of these reagents: (a) the volatility losses of the elements of interest and (b) the contamination of the sample by the usually large volumes of the acids used. In the present official first action AOAC method (2) for the determination of mercury in seafood by flameless atomic absorption, the sample is digested for 3-4 hours with sulfuric, nitric, and perchloric acids, using sodium molybdate as a catalyst (3). The volatility losses of mercury are prevented by use of a condenser partly filled with Raschig rings and glass beads. An efficient device for the removal of perchloric acid fumes is required.

Other techniques for the digestion of organic samples for the determination of volatile elements are available. For example, in arsenic and boron analyses samples are generally ashed at about 500°C. Arsenic losses are prevented by the addition of magnesium nitrate, and boron by the addition of calcium hydroxide.

The volatility losses can be conveniently overcome if the sample is digested in a closed system. Several years ago, Bernas (4) described a specifically designed vessel made of Teflon for difficult-to-decompose inorganic samples (Figure 2). The vessel consists of a Teflon cup which fits snugly inside a stainless steel jacket, that is sealed by a disk of the same material, retained in a stainless steel screw cap. This vessel was used at temperatures up to 170°C for 30 minutes to decompose silicate samples with hydrofluoric acid for subsequent elemental analysis.

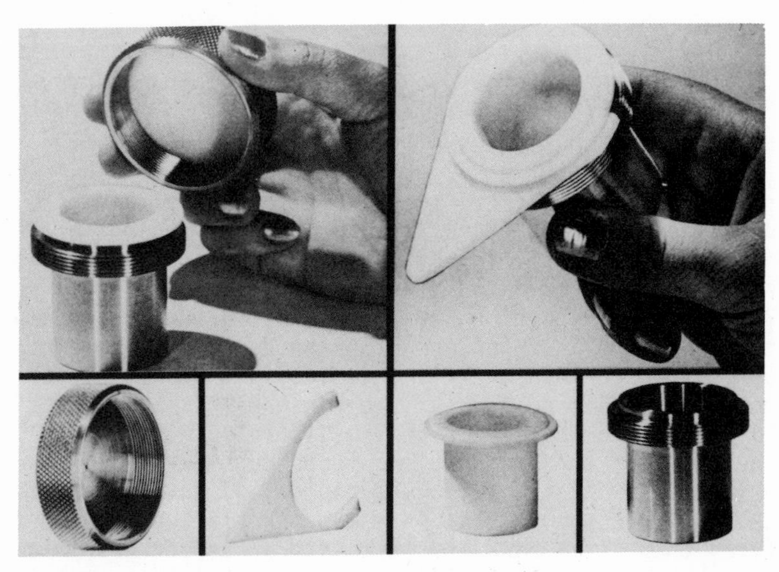

Figure 2. Teflon vessel assembly used for a closed system acid digestion of food samples.

We have substituted nitric acid for hydrofluoric, and have used the same arrangement to digest biological samples for the determination of traces of mercury by flameless atomic absorption (5). The same technique might likewise be used to digest food samples for the determination of other volatile elements including arsenic, selenium and boron.

The procedure for mercury in seafood is as follows: Weigh accurately about 1 g sample (wet basis) into the Teflon vessel. Add 5 ml nitric acid and close the vessel, tightening the screw containing the Teflon sealing disk. Place the vessel into a preheated 150°C oven for 30-60 minutes. Remove the vessel from the oven and allow it to cool to room temperature. Determine the mercury by the Hatch and Ott procedure, as collaborated by Munns and Holland (2).

Recently we conducted a collaborative study in order to validate the above digestion technique. The results of the study were presented at the October, 1973, meeting of the Association of Official Analytical Chemists (6). The method was adopted as "official first action" for mercury in seafood.

GAS-SAMPLING IN ATOMIC ABSORPTION SPECTROPHOTOMETRY

1. The Formation of Gaseous Hydrides

Commercial nebulizers that are used as a means of sample introduction into the flame are only about 5 percent efficient in forming sufficiently fine droplets that can be atomized. In addition, a solution aspirated at the usual rate of 3-5 ml per minute, for say 10 seconds, produces a signal of this duration. If, on the other hand, were the element to be introduced into the flame rapidly with no loss, for example in one second then a larger signal of short duration would result and the detection limit should improve. This is the basis of the gas-sampling techniques in atomic absorption spectrophotometry. The element of interest is converted into a compound that is a gas at relatively low temperature, then it is rapidly swept with a carrier gas into the flame of an atomic absorption spectrophotometer and the absorption signal is recorded.

Elements such as arsenic, selenium, antimony, tellurium, among others react readily with nascent hydrogen to form the respective hydrides. Many devices are now commercially available to carry out the analysis. Essentially, the elements must be in solution in the proper oxidation state, the solution made strongly acidic and zinc granules are added; the nascent hydrogen evolved reacts with the metals producing gaseous hydrides which are then introduced into the atomic absorption spectrophotometer flame. Recently, sodium borohydride has been used in place of zinc as a reducing agent and as a source of hydrogen with some advantages. Orignially, we collected arsine in a liquid nitrogen trap (7) before introducing it into the spectrophotometer flame (Figure 3). Another device commercially available utilizes a rubber balloon as a metal hydride trap (8). Other techniques at the present time rely on the rapid production of hydrides which are then swept into the flame directly.

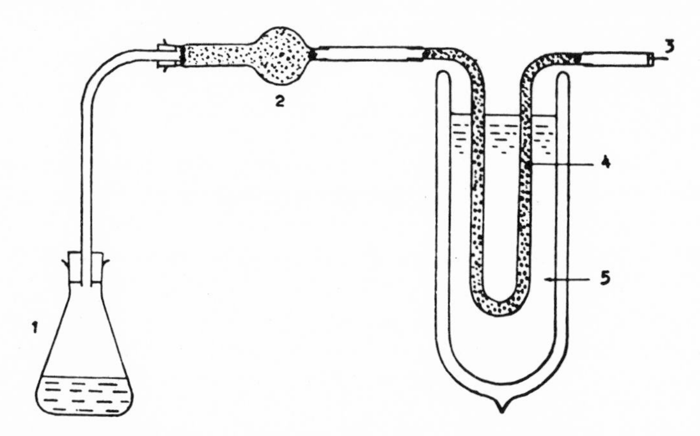

Figure 3. Arsine generator and liquid nitrogen trap. 1) Arsine generator, 2) calcium chloride, 3) needle, 4) glass beads, and 5) liquid nitrogen.

Folger et al. (9) described rapid volatilization of arsenic, selenium, antimony and tellurium in the form of their hydrides for the purpose of studying short-lived nucleids and for the measurement of fission yields. Their results indicate that relatively concentrated hydrochloric acid, at least 6N, is superior to the commonly used sulfuric acid. Violent bursts of hydrogen produced by adding a large surplus of fine-grain zinc powder to a small volume of solution speed up the process. The reagents which resulted in a 80-90% yield of the hydrides in a time period of one second were a combination of 2 ml of conc. HCl and 1.6 g of zinc powder, approximately 240 mesh.

The application of this technique is being investigated in our laboratory for the determination of arsenic and selenium in foods at sub-ppm levels. The digestion of the sample is performed in a closed system using the Teflon vessel, previously described. The procedure is as follows: About 1 g of a representative sample (wet based) is accurately weighed directly into the vessel, 5 ml of nitric acid is added, the vessel is closed and placed in an oven at 150°C for one hour or longer. The sample is transferred to a porcelain crucible, the nitric acid is then evaporated off on a steam bath, magnesium nitrate is added and the residue heated not exceeding 500°C to complete the destruction of the organic matter. Then, the residue is dissolved in conc. HCl and a 2 ml aliquot is used for the determination. The apparatus is illustrated in Figure 4. The detection limit is approximately 0.01 ug for As and Se using a nitrogen-hydrogen-entrained air flame.

2. Formation of Volatile Chelates

The number of elements which readily form hydrides is limited. In order to extend the gas sampling technique to other metals, additional ways must be found to form volatile compounds. Many metals have been determined by gas-liquid chromatography in the form of their volatile chelates. Fluorocarbon beta-diketonate chelates have been shown to be especially well suited for GLC since they are fairly volatile

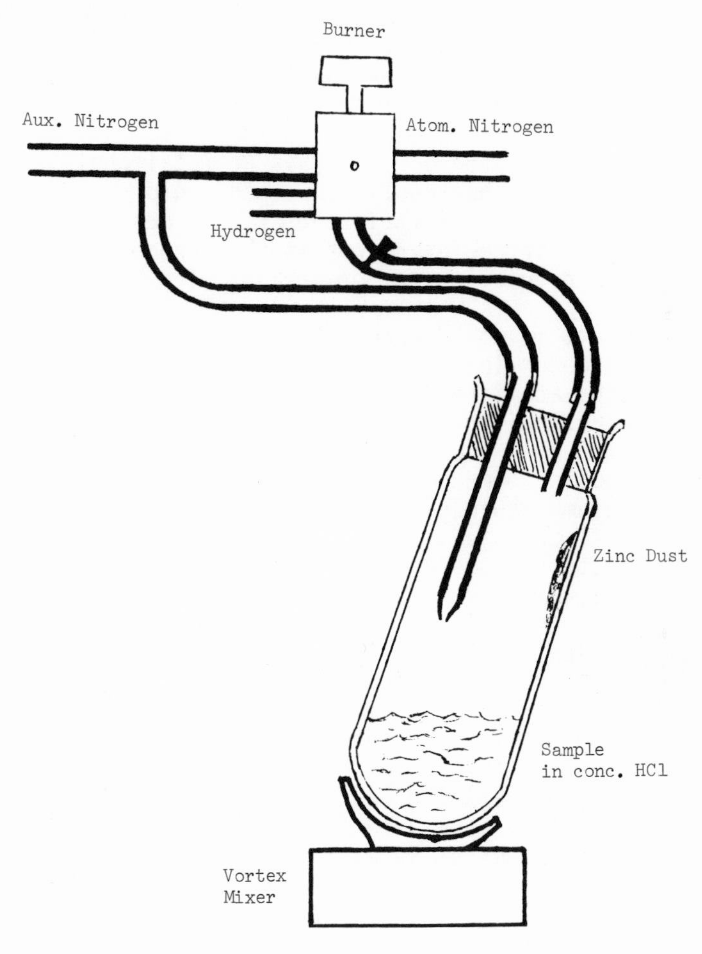

Figure 4. Apparatus for rapid volatilization of hydrides.

at relatively low temperatures and are sufficiently stable so that no decomposition occurs on the column. The most studied complexes are those of 1,1,1-trifluoro-2,3-pentanedione (trifluoroacetylacetone) and 1,1,1,5,5,5-hexafluoro-2,4-pentanedione (hexafluoroacetylacetone). (9) Among the metals which have been successfully gas chromatographed as trifluoroacetylacetonates are: beryllium, aluminum, copper, chromium and iron. The technique is extremely sensitive, especially if an electron capture detector is used.

The success of the GLC analysis of metal chelates depends primarily on the resolution of the column and the complexes are identified by their retention times. Any excess chelating agent often interferes in this process and must be removed by extraction.

In theory, it would appear that the metal chelates which can be determined by GLC should also lend themselves to analysis by atomic absorption using a gas sampling technique. The analysis by AAS would have several advantages over that of GLC. No column to separate the components would be required because AAS is highly specific for the metal. A column would only serve as a source of heat to keep the metal chelates in the gas phase. The atomic absorption spectrophotometer in effect would act as a detector for the metal.

Information in the chemical literature on the direct volatilization of metal chelates for atomic absorption spectrophotometry is scarce. Bailey and Lo (11) used the technique for copper, iron and chromium. They reported an improvement in sensitivity over nebulization, but not as much as expected. They attributed their lack of success to the technique that they have used for the introduction of the sample into the flame of an atomic absorption spectrophotometer.

The technique employed in our laboratory is very similar to that used for conventional GLC. The device is illustrated in Figure 5.

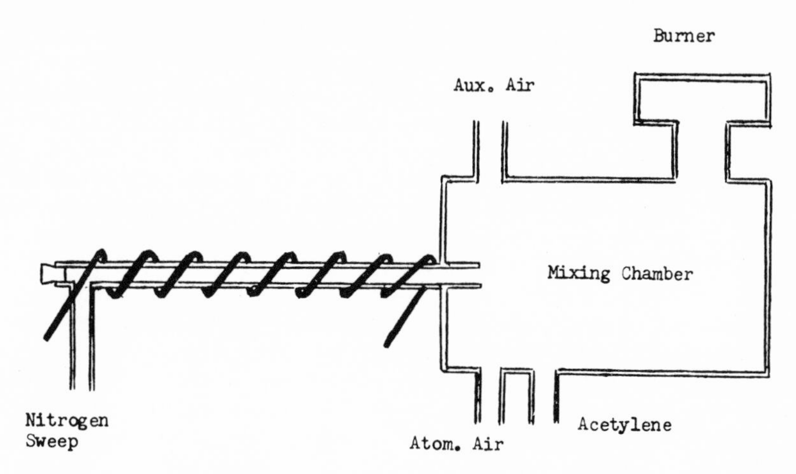

Figure 5. Apparatus for direct volatilization chelates.

Before the technique can be of use in the analysis of foods we must have methods for forming the metal chelates. The most practical approach probably is through extraction into an organic solvent. Scribner et al., (12) give useful extraction schemes for copper, iron, and aluminum by formation of the respective trifluoroacetylacetonates.

The method which we have tried for copper and iron in foods is as follows: Ash about 1 gm of sample at a temperature no higher than 500°C. Dissolve the residue in 0.1 N perchloric acid, adjust the pH with acetate buffer to a value of 5, extract into a solution of trifluoroacetyaceton-ate in methyl isobutyl ketone by shaking for 10 minutes and then letting the phases separate. Inject an appropriate aliquot of the organic layer into the preheated column. Record the resultant metal peak and then use peak heights for quantitation. Figures 6-9 show some examples of the data obtained using this approach.

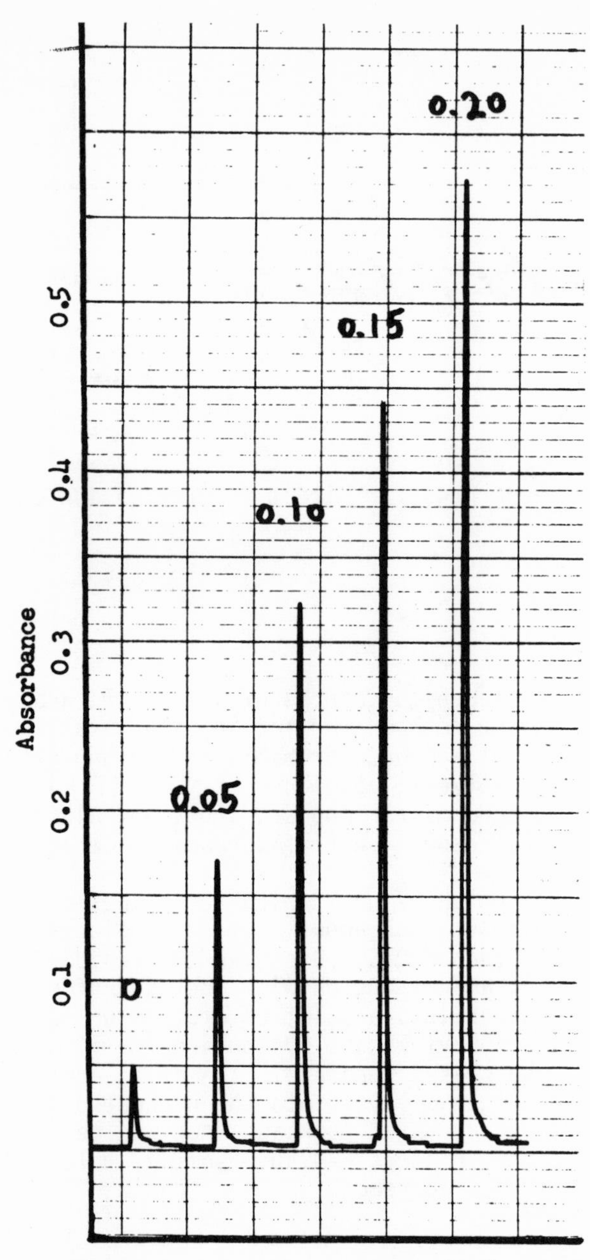

Figure 6. Signals produced by copper (ug), extracted into a solution of trifluoroacetylacetonate in methyl isobutyl ketone, Cu conc. = 0, 5, 10, 15 ug/ml, 10 ul volumes injected.

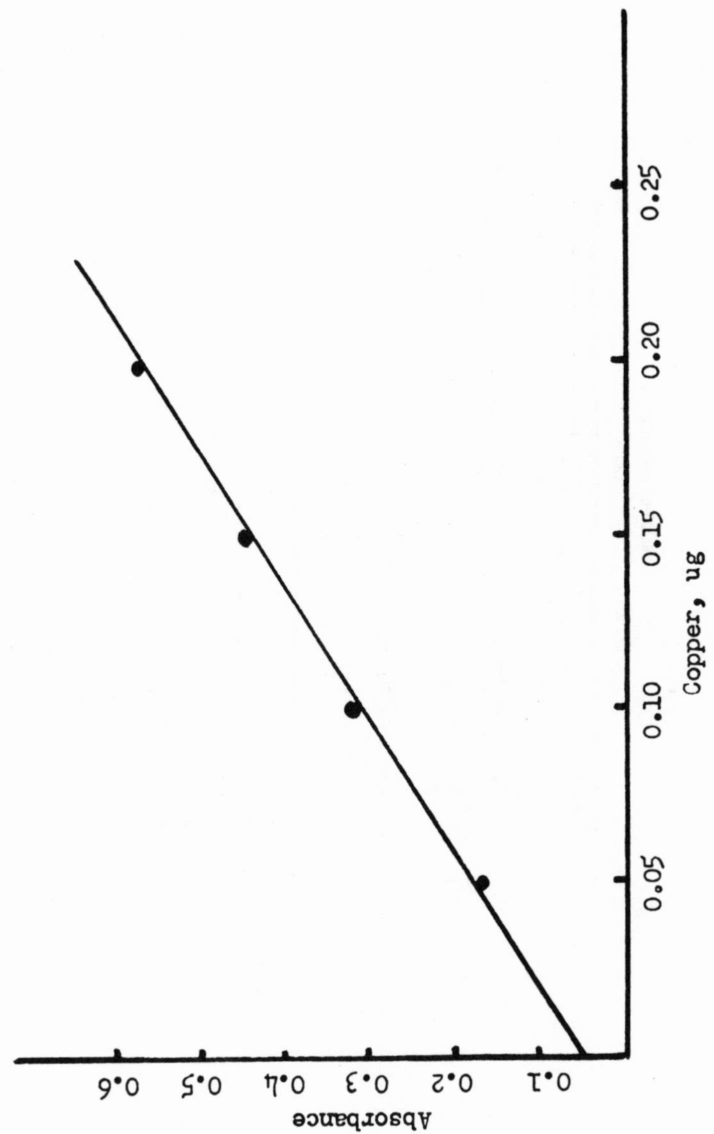

Figure 7. Calibration curve for copper, ug Cu injected vs. absorbance.

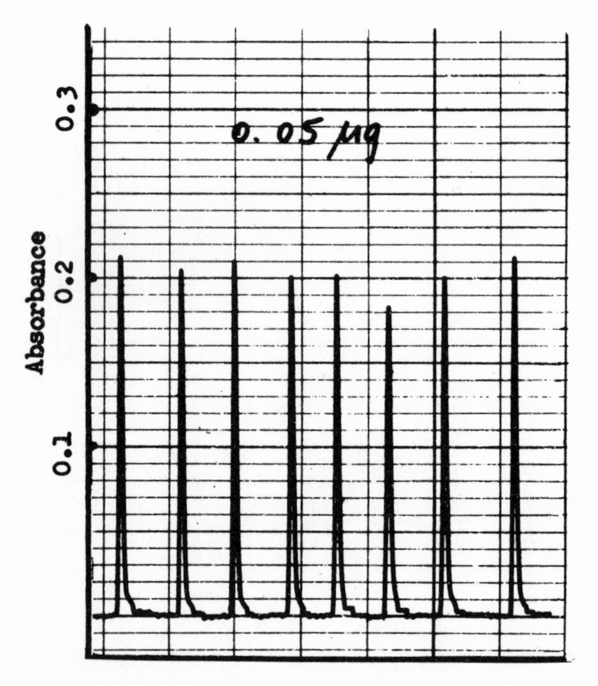

Figure 8. Multiple injections of copper, Cu conc. = 10 ug/ml in methyl isobutyl kentone, 10 ul. volumes injected.

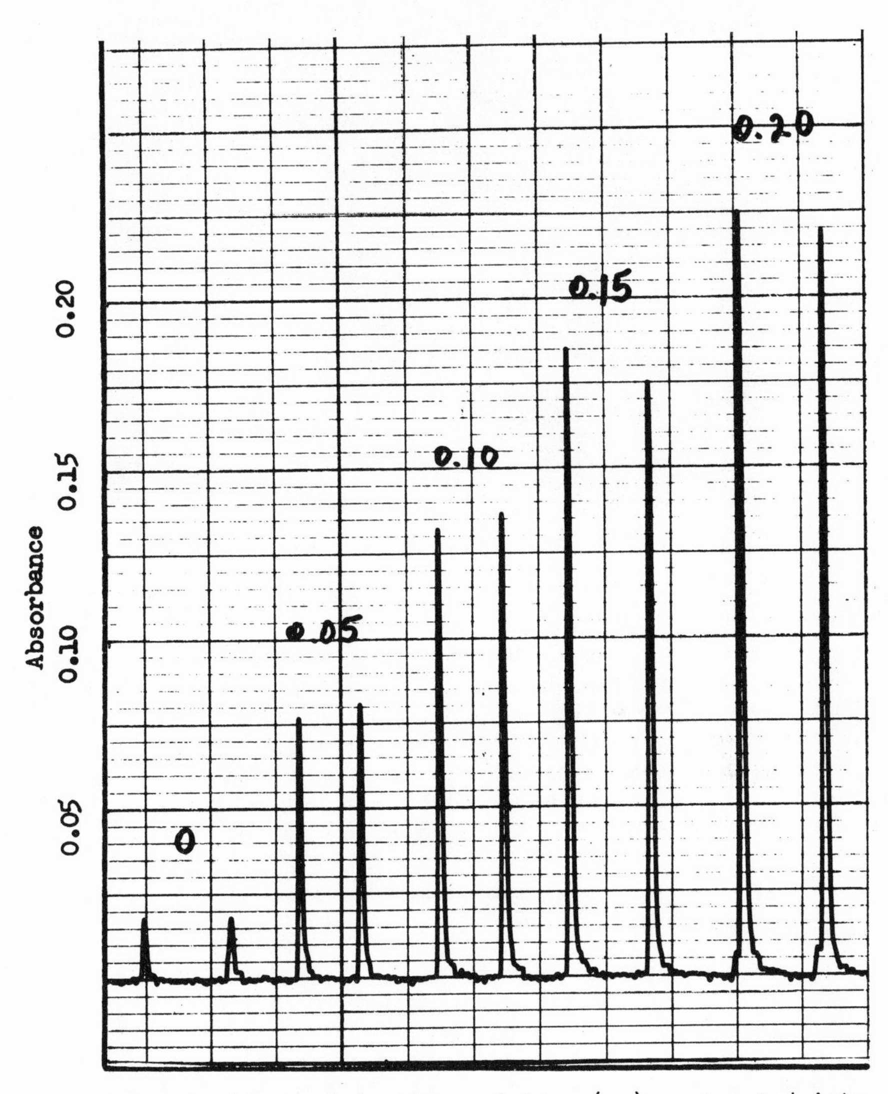

Figure 9. Duplicate injections of iron (ug), extracted into a solution of trifluoroacetylacetonate in methyl isobutyl ketone, Fe conc. = 0, 5, 10, 15 ug/ml, 10 ul volumes injected.

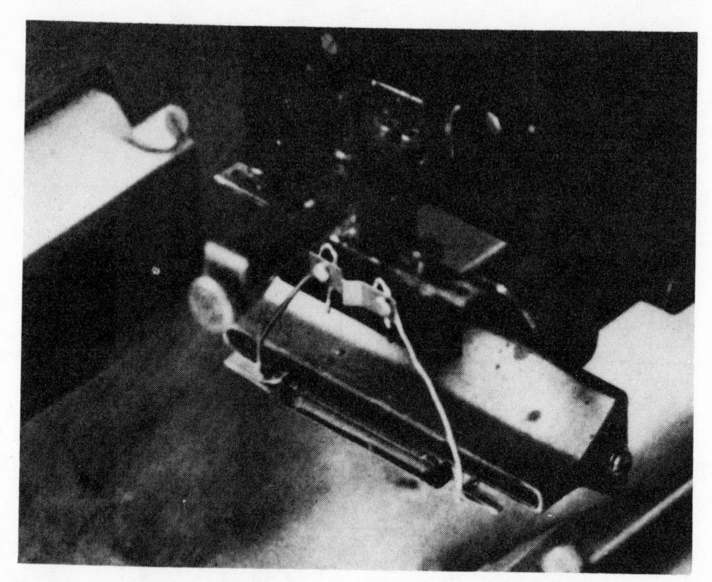

Figure 10. Experimental arrangement using the sampling boat.

One very important advantage of the direct volatilization technique over that of nebulization in addition to an improvement is sensitivity and elimination of many interferences, is the requirement of a much smaller sample volume for the analysis. For example, the sample can be extracted into 1 ml. and since only about 10 ul are used for injection a number of metals can be determined from the single extract.

THE SAMPLING BOAT

Another approach for trace metal analysis, which is gaining in popularity is "flameless" atomic absorption. This is achieved by several commercially available devices, such as the carbon rod, graphite furnace or tantalum ribbon. These devices can accommodate microliter volumes of the sample and may be programmed electrically to dry, ash, and atomize. The detection limits attained by this means of sampling are several orders of magnitude greater than those attained by conventional atomic absorption. However, these devices are still quite expensive and not without drawbacks; for example, matrix interferences, possible non-homogeneity of the small size handled, etc.

We have utilized the tantalum sampling boat for the determination of traces of lead and cadmium in foods (13). Essentially, the difference between the boat and the flameless devices in the preceding paragraph just mentioned is that it can handle as large as 1 ml sample volumes with the air-acetylene flame as the source of heat. In this work we have modified the boat mount to simplify its physical detachment to better suit our requirements. Briefly, the procedure is as follows: Samples of foods are blended with water to a homogeneous mass and then further diluted with water as required. (A Tissuemizer* is ideal for homogenization.) A fraction of a milliliter of the diluted sample is added to the tantalum boat, dried, and ashed at a temperature not higher than 500°C. The boat with ashed sample is then inserted into the air-acetylene flame by means of a slide in order to make the atomic absorption measurement. The manner of calculation is carried out in a simple way. After the sample signal is recorded, a known volume of the standard to approximately match the sample signal intensity is added to the boat and the measurement is repeated. The areas under the peaks or peak heights are compared for quantitation.

The procedure just outlined appears quite attractive in principle. It is simple, very rapid, and requires no reagents which can contaminate the sample. The experimental arrangement is illustrated in Figure 10. Tables 1, 2 and 3 are tabulations of some of the results which we have obtained by the boat method as well as conventional atomic absorption.

The number of elements that can be determined using the sampling boat technique is limited because the temperature that the boat attains in the flame is always less than the temperature of the flame itself. Therefore, only elements whose salts are fairly volatile can be determined. The manufacturer of the sampling boat lists ten metals amenable to this technique. These include, in addition to lead and cadmium, arsenic, bismuth, mercury, selenium, silver, tellurium,

---

*Tekmar Company
P. O. Box 37202
Cincinnati, Ohio 45222

TABLE 2

Lead in Low Acid Foods by Atomic Absorption Spectrophotometry with Improved Modified Sampling Boat Holder

Sample	Found*, ug/g	Std. Dev., ug/g	Added, ug/g	Total Found*, ug/g	% Recovered
Strained baby carrots	0.22	0.07	0.50	0.73	102
Condensed milk	0.32	0.04	0.75	1.05	97
Mushroom soup	0.30	0.07	1.00	1.32	102
Whole kernel corn	0.19	0.04	1.50	1.67	99

* These values represent an average of four individual analyses.

TABLE 3

Cadmium in Low Acid Foods by Atomic Absorption Spectrophotometry with Improved Modified Sampling Boat Holder

Sample	Found*, ug/g	Std. Dev., ug/g	Added, ug/g	Total Found*, ug/g	% Recovered
Strained baby carrots	0.011	0.003	0.050	0.067	112
Condensed milk	0.011	0.002	0.100	0.115	104
Mushroom soup	0.017	0.003	0.050	0.062	90
Whole kernel corn	0.007	-	0.025	0.029	90

* These values represent an average of four individual analyses.

TABLE 4

Cadmium in Strained Baby Carrot Samples

Sample No.	Cadmium Found, ug/g	
	Conventional AAS(2)	Proposed Method
1	0.049	0.063
2	0.072	0.105
3	0.099	0.107
4	0.071	0.068
5	0.070	0.088
6	0.062	0.050
7	0.01	0.011
8	0.096	0.101
9	0.021	0.022
10	0.061	0.046

thallium and zinc. Although antimony is not on the list, our preliminary results indicate that this element can also be determined by the procedure just outlined for lead and cadmium.

Although my first interest relates to metal analysis in foods, an area to which application of similar techniques may be made is hazardous substances, a category no longer the responsibility of FDA. The same boat technique has been used for the determination of lead in paints. The scheme is relatively simple: first, the paint is treated with a suitable solvent such as methyl isobutyl ketone to dissolve the resin and form a suspension. A small amount is then transferred to the boat and the lead is determined as in foods.

In summary, I have presented several techniques which have resulted from a continuing effort to improve the sensitivity, lower the detection limits and eliminate some interferences for metal analysis in atomic absorption spectrophotometry. I wish to emphasize however, that none of the methods outlined can be used at FDA for regulatory purposes until they are validated through subsequent collaborative study.

ACKNOWLEDGMENT

The author is grateful to Dr. Thomas Medwick, Science Advisor, FDA, New York District, and Professor of Pharmaceutical Chemistry, College of Pharmacy, Rutgers University, New Brunswick, New Jersey, for his invaluable assistance in the preparation of this manuscript.

REFERENCES

1. W. R. Hatch, and W. L. Ott, "Determination of Sub-Microgram Quantities of Mercury by Atomic Absorption Spectrophotometry," Anal. Chem., 40 (14) 2085-2087 (1968).

2. "Changes in Methods," J-AOAC 54 (2), pp. 466-467 (1971).

3. R. K. Munns, and D. C. Holland, "Determination of Mercury in Fish by Flameless Atomic Absorption: A Collaborative Study," J-AOAC, 54 (1) 202-205 (1971).

4. B. Bernas, "A New Method for Decomposition and Comprehensive Analysis of Silicates by Atomic Absorption Spectrophotometry," Anal. Chem., 40 (11) 1682-1686 (1968).

5. W. Holak, B. Krinitz, and J. C. Williams, "Simple Rapid Digestion Technique for the Determination of Mercury in Fish by Flameless Atomic Absorption," J-AOAC, 55 (4) 741-742 (1972).

6. B. Krinitz, and W. Holak, "Simple, Rapid Digestion Technique for the Determination of Mercury in Seafood by Flameless Atomic Absorption: A Collaborative Study," J-AOAC, 57 (3).

7. W. Holak, "Gas Sampling Technique for Arsenic Determination by Atomic Absorption Spectrophotometry." Anal. Chem., 41, (12) 1712-1713 (1969).

8. D. C. Manning, "A High Sensitivity Arsenic-Selenium Sampling System for Atomic Absorption Spectroscopy," P. E. Atomic Absorption Newsletter, 10 (6) 123-124 (1971).

9. H. Folger, J. V. Kratz, C. Herrmann, "Rapid Volatilization of Arsenic, Selenium, Antimony, and Tellurium in Form of Their Hydrides," Radiochem. and Radioanalyt. Letters, 1 (3), 185-190 (1969).

10. R. W. Moshier, and R. E. Silvers, Gas Chromatography of Metal Chelates, Pergamon Press, N. J. (1965), 1st Ed.

11. B. W. Bailey, and Fa-Chun Lo, "Direct Volatilization of Inorganic Chelates as a Method of Sample Introduction in Atomic Absorption Spectrometry," Anal. Chem., 44 (7) 1304-1306 (1972).

12. W. G. Scribner, W. J. Treat, J. D. Weiss and R. W. Moshier, "Solvent Extraction of Metal Ions with Trifluoroacetylacetone," Anal. Chem., 37 (9) 1136-1142 (1965).

13. W. Holak, "Determination of Traces of Lead and Cadmium in Foods by Atomic Absorption Spectrophotometry Using the 'Sampling Boat,'" _P.E. Atomic Absorption Newsletter_, _12_ (3) 63-65 (1973).

INDEX